Heat Stress Tolerance in Plants

Heat Stress Tolerance in Plants

Physiological, Molecular and Genetic Perspectives

Edited by
Shabir Hussain Wani
Sher-e-Kashmir University of Agricultural Sciences and Technology of Kashmir, India

Vinay Kumar
Modern College, Savitribai Phule Pune University, India

WILEY

This edition first published 2020

Registered Office(s)
John Wiley & Sons, Inc., 111 River Street, Hoboken, NJ 07030, USA
John Wiley & Sons Ltd, The Atrium, Southern Gate, Chichester, West Sussex, PO19 8SQ, UK

Editorial Office
9600 Garsington Road, Oxford, OX4 2DQ, UK

For details of our global editorial offices, customer services, and more information about Wiley products visit us at www.wiley.com.

Wiley also publishes its books in a variety of electronic formats and by print-on-demand. Some content that appears in standard print versions of this book may not be available in other formats.

Library of Congress Cataloging-in-Publication Data

Names: Wani, Shabir Hussain, editor. | Kumar, Vinay (Biotechnologist), editor.
Title: Heat stress tolerance in plants : physiological, molecular and genetic perspectives / [edited by] Shabir Hussain Wani, Vinay Kumar.
Description: First edition. | Hoboken, NJ : John Wiley & Sons, Inc., 2020. | Includes bibliographical references and index.
Identifiers: LCCN 2019032800 (print) | LCCN 2019032801 (ebook) | ISBN 9781119432364 (cloth) | ISBN 9781119432395 (adobe pdf) | ISBN 9781119432388 (epub)
Subjects: LCSH: Plants--Effect of heat on. | Plants--Effect of stress on. | Vegetation and climate--Research.
Classification: LCC QK755.5 .H43 2020 (print) | LCC QK755.5 (ebook) | DDC 581.4/2--dc23
LC record available at https://lccn.loc.gov/2019032800
LC ebook record available at https://lccn.loc.gov/2019032801

Cover Design: Wiley
Cover Images: Miaze plant damaged ©Aphakorn Fuengtee/Shutterstock, DNA structure isolated © andrea crisante/Shutterstock

Set in 10/12pt WarnockPro by SPi Global, Chennai, India

10 9 8 7 6 5 4 3 2 1

Contents

List of Contributors

Muna A. Ali
Department of Applied Biology, College of Sciences
University of Sharjah
Sharjah,
UAE

Muhammad Amjad Ali
Department of Plant Pathology
University of Agriculture
Faisalabad,
Pakistan

Farrukh Azeem
Department of Bioinformatics and Biotechnology
Government College University Faisalabad (GCUF)
Faisalabad,
Pakistan

and

Haikou Experimental Station
Chinese Academy of Tropical Agricultural Sciences (CATAS), Hainan Key Laboratory of Banana Genetic Improvement
Haikou, Hainan Province
China

Aditya Banerjee
Department of Biotechnology
St. Xavier's College (Autonomous)
Kolkata, West Bengal,
India

Mansi Bhardwaj
ICAR-National Institute for Plant Biotechnology
IARI Campus, New Delhi,
India

Jipsi Chandra
School of Studies in Biotechnology
Pt. Ravishankar Shukla University
Raipur,
India

Vibhuti Chandrakar
School of Studies in Biotechnology
Pt. Ravishankar Shukla University
Raipur,
India

Muhammad Tanees Chaudhary
Department of Plant Breeding and Genetics
University of Agriculture
Faisalabad,
Pakistan

Aksar Ali Chowdhary
Department of Botany
Central University of Jammu
Samba,
India

Sagar Satish Datir
Biology Department
Queen's University
Kingston, ON,
Canada

and

Department of Biotechnology
Savitribai Phule Pune University
Pune, MS,
India

Xiongming Du
Institute of Cotton Research of Chinese Academy of Agricultural Sciences, State Key Laboratory of Cotton Biology
Anyang, Henan,
China

Madhuri Gupta
College of Biotechnology
Sardar Vallabhbhai Patel University of Agriculture & Technology
Meerut, Uttar Pradesh,
India

Mohamed Helmy
The Donnelley Centre
University of Toronto
Toronto, ON,
Canada

Usman Ijaz
Department of Bioinformatics and Biotechnology
Government College University Faisalabad (GCUF)
Faisalabad,
Pakistan

Ahmed Ismail
Department of Biotechnology, Faculty of Agriculture
Al-Azhar University
Cairo,
Egypt

and

Public Health Department,
College of Applied Medical Science
Majmaah University
Majmaah City,
KSA

Tariq Jameel
Department of Plant Breeding and Genetics
University of Agriculture
Faisalabad,
Pakistan

Inès Karmous
Biotechnology Laboratory,
Department of Biology
Bolu Abant Izzet Baysal University
Bolu,
Turkey

Parwinder Kaur
UWA School of Agriculture and Environment
The University of Western Australia
Crawley, Western Australia,
Australia

S. Keshavkant
School of Studies in Biotechnology
Pt. Ravishankar Shukla University
Raipur,
India

and

National Center for Natural Resources
Pt. Ravishankar Shukla University
Raipur,
India

Shivani Khanna
College of Biotechnology
Sardar Vallabhbhai Patel University of Agriculture & Technology
Meerut, Uttar Pradesh,
India

Pankaj Kumar
College of Biotechnology
Sardar Vallabhbhai Patel University of Agriculture & Technology
Meerut, Uttar Pradesh,
India

Vinay Kumar
Department of Biotechnology
Modern College of Arts, Science and Commerce,
Savitribai Phule Pune University, Pune,
India

Reetika Mahajan
Govt. Degree College Poonch
Jammu and Kashmir,
India

Shakti Mehrotra
Department of Biotechnology
Institute of Engineering and Technology
Lucknow,
India

Sonal Mishra
School of Biotechnology
University of Jammu
Jammu,
India

Kareem A. Mosa
Department of Biotechnology,
Faculty of Agriculture
Al-Azhar University
Cairo,
Egypt

and

Department of Applied Biology, College of Sciences
University of Sharjah
Sharjah,
UAE

Habibullah Nadeem
Department of Bioinformatics and Biotechnology
Government College University Faisalabad (GCUF)
Faisalabad,
Pakistan

Koushlesh Ranjan
Department of Veterinary Physiology and Biochemistry
College of Veterinary and Animal Sciences, Sardar Vallabhbhai Patel University of Agriculture and Technology
Meerut, Uttar Pradesh,
India

Aryadeep Roychoudhury
Department of Biotechnology
St. Xavier's College (Autonomous)
Kolkata, West Bengal,
India

Jatindra Kumar Seth
Plant Molecular Biology and 'OMICS' Laboratory
Regional Plant Resource Centre
Bhubaneswar, Odisha,
India

Jitender Singh
College of Biotechnology
Sardar Vallabhbhai Patel University of Agriculture & Technology
Meerut, Uttar Pradesh,
India

Anil Sirohi
College of Biotechnology
Sardar Vallabhbhai Patel
University of Agriculture &
Technology
Meerut, Uttar Pradesh,
India

Vikas Srivastava
Department of Botany
Central University of Jammu
Samba,
India

Giridara Kumar Surabhi
Plant Molecular Biology and
'OMICS' Laboratory
Regional Plant Resource
Centre
Bhubaneswar, Odisha,
India

Lin Tan
Haikou Experimental Station,
Chinese Academy of Tropical
Agricultural Sciences
(CATAS)
Hainan Key Laboratory of
Banana Genetic Improvement
Haikou, Hainan Province,
China

Muhammad Tehseen Azhar
Department of Plant Breeding
and Genetics
University of Agriculture
Faisalabad,
Pakistan

Sandeep Kumar Verma
Biotechnology Laboratory,
Department of Biology
Bolu Abant Izzet Baysal
University
Bolu,
Turkey

Shabir Hussain Wani
Mountain Research Centre
for Field Crops
Khudwani, Sher-e-Kashmir
University of Agricultural
Sciences and Technology of
Kashmir
Jammu and Kashmir,
India

Roseline Xalxo
School of Studies in
Biotechnology
Pt. Ravishankar Shukla
University
Raipur,
India

Annu Yadav
College of Biotechnology
Sardar Vallabhbhai Patel
University of Agriculture &
Technology
Meerut, Uttar Pradesh,
India

Bhumika Yadu
School of Studies in Biotechnology
Pt. Ravishankar Shukla University
Raipur,
India

Sajad Majeed Zargar
Division of Plant Biotechnology
Sher-e-Kashmir University of Agricultural Science and Technology of Kashmir
Shalimar, Srinagar,
Jammu and Kashmir,
India

Foreword

One of the prevalent environmental stresses encountered by plants during their important growth stages is heat stress. Exposure to heat stress for prolonged periods can even result in plant death. Plants can be damaged by either high day or high night temperatures and by either high air or high soil temperatures. Predictions indicate that temperatures will intensify by another 2–6 ° by the climax of this century which will increase the likelihood of more frequent and more severe heat stress in plants, begetting a serious reduction in crop yields. The genetic basis of heat stress adaptation is poorly understood. Conventional breeding methods have met with limited success in improving the heat-stress tolerance of important crop plants through interspecific or intergeneric hybridization. Therefore, it is imperative to accelerate efforts for unraveling the biochemical, physiological, and molecular mechanisms underlying heat-stress tolerance in plants. Heat stress can induce diverse physiological and molecular responses in plants that lead to the disturbance of various cellular metabolic processes, impair membrane stability, cause protein denaturation and thermal aggregation, and consequently affect plant growth and development. The heat stress response is also illustrated by the inhibition of regular transcription and translation, elevated expression of heat shock proteins (HPSs), and induction of heat tolerance. Recent advances in molecular genetics approaches have provided new insights into the plant heat stress response. The QTL-based approaches allow loci to be identified for markers that are linked to heat tolerance. Further, discovery of novel genes and pathways, analyses of expression patterns, and the determination of function of these genes during heat stress adaptation will offer a basis for efficient engineering strategies with the aim to enhance heat stress tolerance in plants.

I am delighted that the edited volume *Heat-Stress Tolerance in Plants: Physiological, Molecular and Genetic Perspectives* is being published. The chapters included in this book are nicely written by scientists and researchers. This book describes recent advances in heat-stress tolerance utilizing genetic and genomic approaches and their application in important agricultural crops. Recent advancements in plant omics and their importance and application in the development of heat-stress-tolerant crops have also been included.

I congratulate the editors for unravelling this volume and hope that this will be a useful reference material for researchers, students, and policymakers.

9th May 2019 *T. Mohapatra*
New Delhi

About the Book

The global population is 7.3 billion and by 2050 it is expected to reach 9.7 billion, while at the same time agricultural productivity is not increasing with the same speed, owing to the mounting environmental constraints as a result of climate change. Heat stress, which can be defined as a period in which temperatures are hot enough for a sufficient period of time to cause irreversible damage to a plant's function or development, is one of the major environmental stresses hampering plant growth and productivity severely. The projections about the increase in global ambient temperatures over the next few decades are expected to increase the intensity of the impact of heat stress on plants.

Owing to limited genetic basis for plant heat stress responses and tolerance besides limited success of conventional breeding approaches, a deeper understanding of the mechanisms underlying plants' responses to heat stress is needed. Recent investigations confirmed that heat stress inhibits transcription and translation, and increases the levels of heat shock proteins (HSPs). During extreme heat stress conditions, signaling pathways leading to apoptotic cell death are also activated. As molecular chaperones, HSPs offer protection to cells against the destructive effects of heat stress and augment survival. The improved expression of HSPs is regulated by heat shock transcription factors (HSFs). Several recent advances in identifying gene(s) and/or pathways and their roles in heat stress responses and tolerance have offered opportunities for researchers to explore these gene(s) for engineering heat-stress tolerance in major crops. Studies involving full genome profiling/sequencing, and mutational and transgenic plant analyses have provided a deep insight of the complex transcriptional mechanism that operates under heat stress.

Through this book, we intend to integrate the contributions from potential plant scientists targeting heat-stress tolerance mechanisms using physiological, biochemical, molecular, and genetic approaches.

About the Editor

Dr. Shabir Hussain Wani is senior Assistant Professor at Mountain Research Centre for Field Crops, Khudwani – 192101, Sher-e-Kashmir University of Agricultural Sciences and Technology of Kashmir, Jammu and Kashmir, India. He received a PhD in plant breeding and genetics on "transgenic rice for abiotic stress tolerance" from the Punjab Agricultural University Ludhiana, India. After obtaining his PhD, he worked as a research associate in the Biotechnology Laboratory, Central Institute of Temperate Horticulture (ICAR), Srinagar, India. He then joined the Krishi Vigyan Kendra (Farm Science Centre) as program coordinator at Senapati, Manipur, India. He teaches courses related to plant breeding, seed science and technology, and stress breeding, and has published more than 100 papers/chapters in journals and books of international and national repute. He served as guest editor and reviews editor for the journal *Frontier in Plant Science* (2015–2018). He has also edited several books on current topics in crop improvement for abiotic stress tolerance published by Springer Nature and CRC Press. His PhD research fetched first prize in the North Zone Competition, at national level, in India. He was awarded Young Scientist Award from the Society for Promotion of Plant Sciences, Jaipur, India, in 2009. He is a fellow of the Society for Plant Research, India. Recently, he also received Young Scientist Award (Agriculture) 2015 from Society for Plant Research, Meerut, India. He also served as visiting Scientist at Department of Plant Soil and Microbial Sciences, Michigan State University, USA under the UGC Raman Post Doctoral Fellowship program.

Currently, he is in charge of Wheat improvement program at MRCFC Khudwani SKAUST Kashmir.

Dr. Vinay Kumar is an Associate Professor at the Postgraduate Department of Biotechnology, Progressive Education Society's Modern College of Arts, Science and Commerce, Ganeshkhind, Pune, India and a Visiting Faculty at the Department of Environmental Sciences, Savitribai Phule University, Pune, India. He obtained his PhD in Biotechnology from Savitribai Phule Pune University (formerly University of Pune) in 2009. For his PhD, he worked on metabolic engineering of rice for improved salinity tolerance. He has published more than 40 peer-reviewed research/review articles, contributed 18 book chapters and edited 4 books including this, published by Springer and Wiley. He is a recipient of the Government of India's Science and Engineering Board, Department of Science and Technology (SERB-DST) Young Scientist Award in 2011. His current research interests include elucidating molecular mechanisms underlying salinity stress responses and tolerance in plants. His research group is engaged in deciphering the molecular mechanisms responsible plant abiotic stress tolerance.

1

Heat Tolerance in Cotton: Morphological, Physiological, and Genetic Perspectives

Muhammad Tehseen Azhar[1], *Shabir Hussain Wani*[2], *Muhammad Tanees Chaudhary*[1], *Tariq Jameel*[1], *Parwinder Kaur*[3], *and Xiongming Du*[4]

[1] *Department of Plant Breeding and Genetics, University of Agriculture, Faisalabad, Pakistan*
[2] *Mountain Research Centre for Field Crops, Khudwani, Sher-e-Kashmir University of Agricultural Sciences and Technology of Kashmir, Jammu and Kashmir, India*
[3] *UWA School of Agriculture and Environment, The University of Western Australia, Crawley, Western Australia, Australia*
[4] *Institute of Cotton Research of Chinese Academy of Agricultural Sciences, State Key Laboratory of Cotton Biology, Anyang, Henan, China*

1.1 Introduction

Cotton belongings to the genus *Gossypium* and four species, namely *G. hirsutum* L., *G. barbadense* L., *G. arboreum* L., and *G. herbaceum* L., that were grown independently for the purpose of textile fiber (Fryxell 1992). Cotton is mainly grown in tropical and sub-tropical areas having temperatures varying from 40 to 45 °C (Ashraf et al. 1994). The phenological development and accumulation of plant biomass depend upon temperature during its growing season (Echer et al. 2014). Cotton sheds its flowers and squares when the temperature reaches 36 °C (Reddy et al. 1991). The plant reproduction, metabolism, and productivity of cotton are adversely affected by high-temperature stress (Demirel et al. 2014). The response of plants to heat stress depends upon the duration and degree of heat stress (Hasanuzzaman et al. 2013). Furthermore, high-temperature stress is closely related to water deficit and can be further exacerbated by limited or unreliable availability of water in cotton-growing

Heat Stress Tolerance in Plants: Physiological, Molecular and Genetic Perspectives, First Edition.
Edited by Shabir Hussain Wani and Vinay Kumar.

areas (Rizhsky et al. 2004). The presence of genetic variation within a species is a prerequisite for a breeding program for the development of genotypes responsive to heat stress (Azhar et al. 2009). All metabolic and biochemical activities of plants require an optimum temperature range, which is termed the thermal kinetic window (TKW). Plant temperature should be within range of the TKW. The plant comes under heat stress if its temperature goes below or above TKW. A TKW of 23.5–32 °C is required for proper plant growth (Sawan et al. 2009). Plant breeders are continually screening cotton germplasm for heat stress due to climate change. Like other physiological traits, seed cotton yield (SCY) is considered a beneficial trait for the development of cotton germplasm against high temperature but it is complex, and affected by climatic conditions, and so requires very specialized breeding programs, but even then one cannot avoid environmental variations (Demİrel et al. 2016). High temperature exerts negative impact due to inhibition of photosynthesis process (DeRidder and Salvucci 2007). The modifications induced by high temperature may be direct, owing to changes in the physiological process, or can be indirect by altering the developmental patterns. For example, developing seeds may be affected by heat stress that may delay the germination or loss of vigor. Ultimately it will lead to reduced emergence and seedling establishment (Wahid et al. 2007). The exposure of cotton to high temperature causes shedding of 35% premature flower and 50–75% of bolls in variable climatic conditions.

1.1.1 Morphological and Physiological Traits

One of the most economical and suitable ways to mitigate the adverse effects of heat stress is to develop or identify heat-tolerant cultivars. Under various heat stress regimes, plants exhibit different kinds of survival mechanism, which include various heat stress avoidance mechanisms, phonological alterations like variable leaf angle, transpirational mechanism, or changes in lipid membrane composition, and various morphological changes (Niinemets 2010). Plants' stomata closing mechanism, reduced water loss, larger stomatal and trichomatous densities and widening xylem vessels are the most common approaches followed by the plant under different conditions of heat stress (Hasanuzzaman et al. 2013).

Research work about the use of various plant and physiological traits is reviewed in the proceeding paragraphs.

1.1.1.1 Seedling and Root Growth

The germination and development of seedlings of crop plants depend upon remaining with the optimum temperature range of 28 to 30 °C. The base temperature is about 12 °C for seed germination and 15.5 °C for seedling growth. Cool temperature between 2 and 4 °C is a major problem in various places in the United States, mainly across the Delta region of Mississippi during germination and initial growth of the seedlings. Genotypic differences have been observed for germination and root development under cool soil temperatures (Mills et al. 2005). While at the time of sowing in northern India, the wind velocity and soil temperature are very high, resulting in a rapid loss of soil moisture (Lather et al. 2001). McMichael and Burke (1994) reveal that soil with a temperature range of between 20 and 32 °C is suitable for proper root growth and development. Root temperature stress ranges between 35 and 40 °C and badly affects the hydraulic conductivity and nutrient uptake ability of the plant and causes low hormone synthesis, and badly affects hormone transportation (Clark and Reinhard 1991; Burke and Upchurch 1995). Synthesis of cytokinins which originate predominantly in roots is among the most sensitive processes (Paulsen 1994). Many functions of roots – comprising the uptake of nutrients and water, assimilation and synthesis of metabolites, and translocation – are very sensitive to temperature. Root temperature may be more critical than shoot temperatures for plant growth because roots tolerate a shorter range of temperature and are less adaptable to extreme variations (Nielsen 1974).

1.1.1.2 Stomatal Conductance

Stomatal closures to reduce transpiration rates also reduce photosynthesis rates. Increased stomatal conductance helps with the cooling of leaves by evaporation and thereby reduces thermal stress. Optimum temperatures of < 30 °C during daytime in upland and Pima cotton are well below commonly occurring air temperatures in most cotton-growing areas (Radin et al. 1994). Stomatal conductance and net photosynthesis are inhibited by moderate heat stress in many plant species, owing to decreases in the activation state of rubisco (Crafts-Brander and Salvucci 2002; Morales et al. 2003). Stomata could be explored for the development of heat tolerance in crop plants because they regulate the transpiration rate and determine the degree to which water evaporation can cool down the leaves. If stomata open wider in higher-yielding lines, transpiration may be increased and photosynthetic rates improved because CO_2 as well as water vapors must diffuse through the stomata. A broader examination of 16 species performed by Monteith (1995) supported the theory that stomata respond not to humidity levels but to respiration rates. Moneith (1995) also identified that stomata from all investigated species had patchy stomata closure in response to increased respiration rates to limit water loss. Many studies were intended to assess the germplasm of cotton and wheat by using stomatal conductance and photosynthesis as criteria for the identification of heat-tolerant genotypes (Cornish et al. 1991; Lu et al. 1998; Ulloa et al. 2000; Rahman 2005).

1.1.1.3 Cell Membrane Thermostability

The conventional screening methods of large germplasm are time consuming and laborious but the success of any breeding program depends on effective evaluation techniques (Asha and Lal Ahamed 2013). Cell membrane thermostability (CMT) protocol has been suggested by Sullivan (1972) as a dependable method for measuring heat tolerance by quantifying the amount of electrolyte leakage from leaf disks after exposure to heat stress, and this technique has been used efficiently in potato (Coria et al. 1998), rice (Maavimani et al. 2014), tomato (Golam et al. 2012), soybean (Martineau et al. 1979), wheat (Blum et al. 2001), sorghum (Marcum 1998), and barley (Wahid and Shabbir 2005). Membrane thermostability has been reported to have a strong genetic correlation with grain yield in wheat. Heritability of membrane thermostability in maize was estimated to be 73%. CMT has been used in cotton as a suitable screening and selection criterion for heat tolerance, owing to its ability to distinguish heat-sensitive and -tolerant genotypes within the species (Singh et al. 2007). Marsh et al. (2000) found a large portion of the variability for membrane stability to be controlled by a small number of genes. Rahman et al. (2004) indicate that exposure to high temperature preceding the CMT test produces better distinction between heat-tolerant and heat-susceptible cultivars; however, they caution regarding its indirect selection on the basis of SCY under non-heat-stressed environments, so Rahman et al. (2004) conclude that CMT could be a useful physiological technique for screening large sizes of germplasm of cotton for heat tolerance. In addition, chlorophyll fluorescence, membrane leakage, and leaf extension growth (Jamieson et al. 1995) were also the most practical, dependable, but sensitive methods for quantifying the response of cottons to high-temperature stress. This permeability activates a series of calcium-dependent protein kinases (CDPKs) that spur cellular production of increased antioxidants (Wahid et al. 2007). The effectiveness of CMT for field-grown leaf material to determine difference in genotypes in heat tolerance may be very useful in detecting the desirable genotypes for higher temperatures (Singh et al. 2007). However, membrane stability requires destructive sampling, and there is a large potential for error inherent in the process of estimating membrane stability. Similarly, measurements of chlorophyll fluorescence require the use of expensive instrumentation and, in some cases, necessitate dark adaptation of the leaf tissue, which limits the number of plants that can be screened in a given day. The ability of cotton to adjust its membrane structure under heat-stressed conditions has been recognized as a significant physiological adaptation to heat stress (Rahman et al. 2004).

1.1.1.4 Canopy Temperature

Owing to high temperature, nearly 65–70% of fruiting points shed down because of heat-induced pollen sterility, which causes a poor yield of seed cotton at harvest time (Taha et al. 1981; Hofmann et al. 1988; Baloch et al. 2000; Liu et al. 2006). The relationship of canopy temperature (Tc) with transpiration has been used for the detection of heat stress in crop plants (Conaty et al. 2015). The development of heat tolerance in cultivars through conventional breeding approach depends upon the accessibility of an optimized protocol which could give advantage to screen large number of varieties/lines (germplasm) rapidly. A new physiological infrared technique that measures the canopy temperature is becoming increasingly popular these days as a way of screening large amount of germplasm for heat tolerance (Singh et al. 2007). This technique had been used by various workers in cotton and corn for the identification of heat-tolerant lines which were used independently in breeding programs (Wanjura et al. 2004; Karademir et al. 2012). The same assay was used for drought tolerance in cotton and wheat (Hatfield et al. 1987; Rashid et al. 1999; Blum et al. 2001). The use of canopy temperature to assess plant growth and development where there is less irrigation may be more reliable than air temperature alone (Mahan et al. 2014). Previous studies have evaluated the optimum thermal environment for growth and metabolism, placing the thermal optimum for cotton at 28 °C, within a range of 25–31 °C (Burke 1990; Hatfield and Burke 1991; Mahan et al. 1995; Burke et al. 2004; Conaty et al. 2012). The PAR model is used to detect the canopy temperature with optimal rate of photosynthesis to produce different consequences of plant architecture (Alarcon and Sassenrath 2015).

Crop management practices including suitable plant population density (PPD) are an important technique for adjusting canopy light distribution and increasing canopy photosynthetic capacity of cotton grown in the field (Yao et al. 2016). Enhancements in canopy photosynthetic capacity are essential to increase yield (Zhang et al. 2004; Dong et al. 2012; Mao et al. 2014; Dai et al. 2015). Canopy photosynthetic capacity is also significantly affected by proper crop management practices (Dong and Liu 2010; Dong et al. 2012; Mao et al. 2014; Dai et al. 2015).

1.1.1.5 Chlorophyll Content

Photosynthesis is extremely sensitive to the effect of increased temperatures. Reduction in the photosynthetic rate has been constantly shown in cotton, and is due to the decline in available carbohydrate supply to the developing bolls, leading to increased rates of fruit shed (Guinn 1976; Hake and Silvertooth 1990; Sadras 1995; Brown and Oosterhuis 2004). Hall and Allen (1993) hypothesize that cultivars displaying heat tolerance during reproductive development should have the following characteristics: a high harvest index, high chlorophyll content per unit leaf area, small leaves, and low leaf area per unit ground area. Heat stress decreases photosynthesis, quantum efficiency, and the chlorophyll content of plants, but increases stomatal conductance. Photosynthesis is an important process that is detrimentally affected by adverse environmental conditions such as high temperature, drought, and salinity, etc. (Cothren 1999). Cotton grown under increased CO_2 and higher temperatures has greater leaf area, increased photosynthetic capacity, and a higher boll retention when compared to normal CO_2 and higher temperatures (Radin et al. 1987; Mauney et al. 1994). Heat imposes negative impacts, such as reduced leaf water potential, reduced leaf area, and premature leaf senescence, which has negative impacts on total photosynthesis performance of plant (Greer and Weedon 2013). Temperatures beyond the optimal thermal limits of the species will hinder the photosynthetic capability (Bibi et al. 2008). This is the result of photosynthesis transitioning from noncyclic photophosphorylation pathways to a cyclic phosphorylation pathway. During the vegetative stage, high temperatures disturb the physiological mechanisms in the leaves thereby decreasing CO_2 assimilation rates, damaging components of photosystem II (PSII) in thylakoid membranes of the chloroplasts, and causing membrane damage (Hall 2004). Cotton maintained significantly higher rates of photosynthesis at 36 °C with no reduction in Rubisco (ribulose-1,5-bisphosphate carboxylase/oxygenase) activase activity at CO_2 concentrations of 450 parts per million (ppm), compared to plants at the same temperature but with lower CO_2 concentrations. Chlorophyll content has been used to assess high temperature tolerance in *Zea mays* L. (Karim et al. 2000) and *Triticale* (Grzesiak et al. 2003); however, use of chlorophyll content as a screening technique for high temperature tolerance in cotton is rare. But genotypic differences in cottons for chlorophyll contents have been identified by various workers (Pettigrew et al. 1993; Pettigrew and Meredith 1994; Pettigrew and Meredith Jr 2012; Quisenberry et al. 1994; Pettigrew and Turley 1998; Clement et al. 2013). These genotypic differences and efficacy of the use of technique allow the cotton breeder to initiate breeding work for the development of heat-tolerant lines that enhance yield and fiber quality traits.

1.1.2 Genetics and Molecular Basis of Heat Tolerance in Cotton

High-throughput molecular technologies assisted with advanced phenotyping platforms provide new opportunities for discovering marker-trait associations and can sustain genomic-assisted breeding (Barabaschi et al. 2016). Understanding the genetic basis of traits related to heat tolerance is essential for the development of heat-tolerant genotypes. Basically, heat tolerance is not associated with a single trait; rather, it is a complex and polygenic mechanism in field crops. Several morphological and physiological characteristics – such as sterility of pollens, electrolyte leakage, CMT, chlorophyll fluorescence, and heat shock proteins (HSPs) in cotton, sorghum, wheat, and maize – have been explored extensively (Joshi 1999). Heat-tolerance responses have also been studied in *Arabidopsis thaliana* (Scharf et al. 2001), *Helianthus annuus* L. (Coca et al. 1996), *Nicotiana tabacum* L. (Rizhsky et al. 2002), and *Solanum tuberosum* L. (Havaux 1993).

Plants respond to heat stress within 0.5–1 hour after exposure to high temperature but it may vary in range of species, by accumulating certain proteins like HSPs, moreover this production of HSPs also depends on the temperature that the plant experiences (Nakamoto and Hiyama 1999; Schoffl et al. 1999; Cottee et al. 2014). These HSPs and heat shock transcription factors are synthesized by genes encoding heat stress. While the rate of respiration increases, this may be due to catabolism of assimilates and glycolysis, and it can also indicate the up-regulation of stress-responsive genes. Besides, genes associated with transport, signaling, and metabolism of sulfur and nitrogen also enhance under stress (Cottee et al. 2014). Burke and Chen (2015) report eight HSPs produced in the dry land cotton leaves. The production of polypeptide was quantified in cotton grown under control and heat-shocked environment, and 11 different polypeptides were produced in a leaf sample of heat shocked cotton, and eight of these HSPs were similar to proteins accumulated in dry land cotton (Burke et al. 1985). Demirel et al. (2014) studied 25 ESTs (express sequence tags) out of which 16 were homologous to known genes. The genes, namely *TH1*, *FPGS*, *GhHS126*, *GhHS128*, *and IAR3* were up-regulated, while *CTL2, RPS14, ABCC3, LSm8,* and *CIPK* expression was down-regulated but expressions of *PP2C* and psaB-*rps14* was not altered, owing to short-term heat stress in cotton. The mechanism of HSPs under high temperature is still unknown. Different studies show that HSPs are molecular chaperones that have a role in protecting cell proteins from denaturation under high temperature (Bowen et al. 2002).

1.1.3 Conventional Breeding Approaches

Traditional breeding of heat-resistant crops basically based on selection and common technique of selecting crops for heat tolerance has been to grow breeding materials in a hot target production environment and detect individuals/lines with a higher yield (Ehlers and Hall 1998). Development of cotton lines by conventional breeding methods reduces the loss of yield reported due to heat stress. Selection of breeding lines for high temperature always takes place in hot regions at critical stages of crop plants (Mickelbart et al. 2015; Driedonks et al. 2016). The researchers report that the optimum temperature of 23 °C is required for carbon fixation during photosynthesis and temperatures above this may affect the photosynthetic processes of cotton plant and adversely affect the yield of seed cotton and fiber characteristics (Bibi et al. 2008; Song et al. 2014). Song et al. (2014) reveal that high temperature affect the rate of pollen germination and boll setting. Numbers of potential heat-tolerant lines of cotton were identified as having viable pollens after the exposure of their pollens at 35 °C for 15 minutes. The same assay was used for the identification of heat-tolerant families during segregating populations. In addition, Abro et al. (2015) identified heat-tolerant genotypes based on relative cell injury percentage (RCI %), heat susceptible index (HSI) value, and boll retention percentage on individual genotypes. These are simple, reliable, and inexpensive for screening methods to heat tolerance. Development of heat resistance needs more attention in comparison to other abiotic stresses. The information on heat tolerance in cotton is scanty in the literature. Even in the presence of all of complications in the mechanisms of heat resistance, several heat-resistant inbred lines and hybrid varieties of tomato have been improved and released for commercial cultivation (Scott et al. 1995, 1996).

1.1.4 Modern Molecular Breeding Approaches

Conventional breeding is tedious, time-consuming, and dependent on the environment as compared to molecular breeding, particularly marker-assisted selection, which is highly efficient and precise. Biotechnology has contributed significantly in understanding and improving for high temperature tolerance in crop plants. Various approaches – namely molecular markers, correlation, co-segregation, and genetic variability – were being used to understand genetic bases of thermotolerance (Maestri et al. 2002). In order to cope with heat stress, plants use various mechanisms, including maintenance of membrane stability, scavenging of reactive oxygen species (ROS), production of antioxidants, accumulation and adjustment of compatible solutes, induction of mitogen-activated protein kinase (MAPK) and CDPK cascades, and most importantly chaperone signaling and transcriptional activation. All these mechanisms, which are regulated at the molecular level, enable plants to thrive under heat stress. Based on understanding such mechanisms, potential genetic strategies to improve plant heat-stress tolerance include traditional and contemporary molecular breeding protocols and transgenic approaches (Wahid et al. 2007). With the advent of DNA marker technique in the 1980s (Botstein et al. 1980) and early 1990s, lot of restrictions linked to isozyme and morphological markers were conquered, and genetic mapping entered a new and exciting era which promised to greatly enhance the efficiency of crop breeding and genetics study. Hall (2001) suggests that genetic engineering is a useful tool for ideotype breeding, because a single gene can be inserted without changing the background of cultivars, and successful stories abound in the literature concerning the development of isogenic lines in a short period. In isogenic lines, differential expression of an inserted gene can be evaluated in a range of environments very precisely. Isogenic lines developed by traditional backcrossing methods require several years to grow generations as compared to marker assisted selection (MAS). It is an efficient approach for the improvement of a stress-tolerant crop. This approach requires the availability of genetic markers linked with stress-tolerant genes. Genetic markers associated in different environmental stresses such as drought, heat, and salinity have been identified in different crop species (Foolad 2005). For example, four quantitative trait loci (QTLs) for heat tolerance were identified in *A. thaliana* (Hong and Vierling 2000). Lin et al. (1996) detected 14 random amplified polymorphic DNA (RAPD) markers associated with heat-resistant characteristics in tomatoes under heat stress with the use of the bulked segregant analysis. In reaction to high temperatures, special HSPs have been distinguished in various crop species. Experimental data gained from transgenic, reverse-genetics, and mutation attributes in noncereal species prove causal involvement of HSPs in thermoresistance in crops (Queitsch et al. 2000). For instance, HSP68, that is located in mitochondria and usually expressed incorporately, was detected to have enhanced expression under heat stress in barley, tomato, maize, potato, and soybean cells (Newman et al. 1993). Likewise genotyping by sequencing (GBS) is also a cost-effective method to detect and score thousands of genome-wide single nucleotide polymorphism (SNP) from diverse populations of crop plants (Elshire et al. 2011; Poland and Rife 2012). In the same way, 11 QTLs were identified in maize by using genetic marker RFLP (restriction fragment length polymorphism) responsible for the growth of pollen tube and pollen germination in heat-stress conditions (Frova and Sari-Gorla 1994). Microarray technology has recently become a powerful tool for the systematic analysis of expression (or transcriptome) profiles of large numbers of genes that are induced or repressed by heat treatment. An initial study in *N. tabacum* examined the transcripts of 170 cDNAs from plants drought-stressed with or without simultaneous heat stress. In current scenarios, several techniques are available to plant scientists that could be utilized for the development of new germplasm that could cope with stressful environments, and ensure food security, because many changes have already been made to the behavior of crops plants which lead to reduced yields.

1.2 Conclusion and Future Prospects

Heat stress has become a major and wide problem for crop production because it affects the development, growth, and productivity of crop plants. However, the degree of occurrence of heat stress varies greatly among climatic zones and depends on the period and probability of high temperature and on the timing of diurnal plants during high temperature. The global ambient temperature is increasing day by day, owing to rapid and increased emission of greenhouse gases including NO_2 and CO_2 from different industries and automobile sources resulting in global warming. Therefore, plant adaptations and responses against enhanced temperatures and the mechanisms which are responsible for the identification or development of heat-tolerant cultivars must be well understood for important agricultural crops. Plant responses to high temperatures also vary across and within species, as well as at different developmental stages. Hence the development of more heat-tolerant cotton has been a primary focus for breeders and geneticists, and researchers examining heat-stress-related effects with chemical ameliorants. Development of heat-tolerant cotton is limited in that the domestication of wild type cotton into its current lineages has reduced the genetic diversity for thermotolerance.

References

Abro, S., Rajput, M.T., Khan, M.A. et al. (2015). Screening of cotton (*Gossypium hirsutum* L.) genotypes for heat tolerance. *Pakistan Journal of Botany* 47 (6): 2085–2091.

Alarcon, V.J. and Sassenrath, G.F. (2015). Optimizing canopy photosynthetic rate through PAR modeling in cotton (*Gossypium spp.*) crops. *Computers and Electronics in Agriculture* 119: 142–152.

Asha, R. and Lal Ahamed, M. (2013). Screening of cotton genotypes for heat tolerance. *International Journal of Bio-resource and Stress Management* 4: 599–604.

Ashraf, M., Saeed, M.M., and Qureshi, M.J. (1994). Tolerance to high temperature in cotton (*Gossypium hirsutum* L.) at initial growth stages. *Environmental and Experimental Botany* 34 (3): 275–283.

Azhar, F.M., Ali, Z., Akhtar, M.M. et al. (2009). Genetic variability of heat tolerance, and its effect on yield and fibre quality traits in upland cotton (*Gossypium hirsutum* L.). *Plant Breeding* 128: 356–362.

Baloch, M.J., Lakho, A.R., Bhutto, H.U. et al. (2000). Estimates of combining ability and genetic parameters for yield and fiber traits in upland cotton. *Pakistan Journal of Biological Sciences* 3: 1183–1186.

Barabaschi, D., Tondelli, A., Desiderio, F. et al. (2016). Next generation breeding. *Plant Science* 242: 3–13.

Bibi, A.C., Oosterhuis, D.M., and Gonias, E.D. (2008). Photosynthesis, quantum yield of photosystem II and membrane leakage as affected by high temperatures in cotton genotypes. *Journal of Cotton Science* 12: 150–159.

Blum, A., Klueva, N., and Nguyen, H.T. (2001). Wheat cellular thermo-tolerance is related to yield under heat stress. *Euphytica* 117: 117–123.

Botstein, D., R.L. White, M. Skolnick and R.W. Davis. 1980. Construction of a genetic linkage map in man using restriction fragment length polymorphisms. American journal of human genetics. 32:314.

Bowen, J., Lay-Yee, M., Plummer, K.I.M., and Ferguson, I.A.N. (2002). The heat shock response is involved in thermotolerance in suspension-cultured apple fruit cells. *Journal of Plant Physiology* 159: 599–606.

Brown, R.S. and Oosterhuis, D.M. (2004). High daytime temperature stress effects on the physiology of modern versus obsolete cotton cultivars. *Summaries of Cotton Research*: 63–67.

Burke, J.J. (1990). Variation among species in the temperature dependence of the reappearance of variable fluorescence following illumination. *Plant Physiology* 93: 652–656.

Burke, J.J. and Chen, J. (2015). Enhancement of reproductive heat tolerance in plants. *PLoS One* 10: e0122933.

Burke, J.J. and Upchurch, D.R. (1995). Cotton rooting patterns in relation to soil temperatures and the thermal kinetic window. *Agronomy Journal* 87: 1210–1216.

Burke, J.J., Hatfield, J.L., Klein, R.R., and Mullet, J.E. (1985). Accumulation of heat shock proteins in field-grown cotton. *Plant Physiology.* 78: 394–398.

Burke, J.J., Velten, J., and Oliver, M.J. (2004). In vitro analysis of cotton pollen germination. *Agronomy Journal* 96: 359–368.

Clark, R.B. and Reinhard, N. (1991). Effects of soil temperature on root and shoot growth traits and iron deficiency chlorosis in sorghum genotypes grown on a low iron calcareous soil. *Plant and Soil* 130: 97–103.

Clement, J.D., Constable, G.A., and Conaty, W.C. (2013). CO_2 exchange rate in cotton does not explain negative associations between lint yield and fiber quality. *Journal of Cotton Science* 17: 270–278.

Coca, M.A., Almoguera, C., Thomas, T.L., and Jordano, J. (1996). Differential regulation of small heat-shock genes in plants: analysis of a water-stress-inducible and developmentally activated sunflower promoter. *Plant Molecular Biology* 31: 863–876.

Conaty, W.C., Burke, J.J., Mahan, J.R. et al. (2012). Determining the optimum plant temperature of cotton physiology and yield to improve plant-based irrigation scheduling. *Crop Science* 52: 1828–1836.

Conaty, W.C., Mahan, J.R., Neilsen, J.E. et al. (2015). The relationship between cotton canopy temperature and yield, fibre quality and water-use efficiency. *Field Crops Research* 183: 329–341.

Coria, N.A., Sarquís, J.I., Peñalosa, I., and Urzúa, M. (1998). Heat-induced damage in potato (*Solanum tuberosum*) tubers: membrane stability, tissue viability, and accumulation of glycoalkaloids. *Journal of Agricultural and Food Chemistry* 46: 4524–4528.

Cornish, K., Radin, J.W., Turcotte, E.L. et al. (1991). Enhanced photosynthesis and stomatal conductance of Pima cotton (*Gossypium barbadense* L.) bred for increased yield. *Plant Physiology* 97: 484–489.

Cothren, J.T. (ed.) (1999). Physiology of the cotton plant. In: *Cotton: Origin, History, Technology, and Production*, 207–268. Wiley Blackwell.

Cottee, N.S., Wilson, I.W., Tan, D.K., and Bange, M.P. (2014). Understanding the molecular events underpinning cultivar differences in the physiological performance and heat tolerance of cotton (*Gossypium hirsutum*). *Functional Plant Biology* 41: 56–67.

Crafts-Brander, S.J. and Salvucci, M.E. (2002). Sensitivity of photosynthesis in a C4 plant, maize, to heat stress. *Plant Physiology* 129: 1773–1780.

Dai, J., Li, W., Tang, W. et al. (2015). Manipulation of dry matter accumulation and partitioning with plant density in relation to yield stability of cotton under intensive management. *Field Crops Research* 180: 207–215.

Demirel, U., Gür, A., Can, N., and Memon, A.R. (2014). Identification of heat responsive genes in cotton. *Biologia Plantarum* 58: 515–523.

Demİrel, U., Çopur, O., and Gür, A. (2016). Early-stage screening for heat tolerance in cotton. *Plant Breeding* 135: 80–89.

DeRidder, B.P. and Salvucci, M.E. (2007). Modulation of Rubisco activase gene expression during heat stress in cotton (*Gossypium hirsutum* L.) involves post-transcriptional mechanisms. *Plant Science* 172: 246–254.

Dong, K. and Liu, Y. (2010). Cross-cultural management in China. *Cross Cultural Management: An International Journal* 17: 223–243.

Dong, H., Li, W., Eneji, A.E., and Zhang, D. (2012). Nitrogen rate and plant density effects on yield and late-season leaf senescence of cotton raised on a saline field. *Field Crops Research* 126: 137–144.

Driedonks, N., Rieu, I., and Vriezen, W.H. (2016). Breeding for plant heat tolerance at vegetative and reproductive stages. *Plant Reproduction* 29: 67–79.

Echer, F.R., Oosterhuis, D.M., Loka, D.A., and Rosolem, C.A. (2014). High night temperatures during the floral bud stage increase the abscission of reproductive structures in cotton. *Journal of Agronomy and Crop Science* 200: 191–198.

Ehlers, J.D. and Hall, A.E. (1998). Heat tolerance of contrasting cowpea lines in short and long days. *Field Crops Research* 55: 11–21.

Elshire, R.J., Glaubitz, J.C., Sun, Q. et al. (2011). A robust, simple genotyping-by-sequencing (GBS) approach for high diversity species. *PLoS One* 6: 19–79.

Foolad, M.R. (2005). Breeding for abiotic stress tolerances in tomato. In: *Abiotic Stresses: Plant Resistance Through Breeding and Molecular Approaches* (ed. P. Harris), 613–684. New York: The Haworth press Inc.

Frova, C. and Sari-Gorla, M. (1994). Quantitative trait loci (QTLs) for pollen thermotolerance detected in maize. *Molecular and General Genetics* 245: 424–430.

Fryxell, P.A. (1992). A revised taxonomic interpretation of *Gossypium* L. (Malvaceae). *Rheedea* 2: 108–116.

Golam, F., Prodhan, Z.H., Nezhadahmadi, A., and Rahman, M. (2012). Heat tolerance in tomato. *Life Science Journal* 9: 1936–1950.

Greer, D.H. and Weedon, M.M. (2013). The impact of high temperatures on *Vitis vinifera* cv. Semillon grapevine performance and berry ripening. *Frontiers in Plant Science*: 4.

Grzesiak, S., Grzesiak, M.T., Filek, W., and Stabryła, J. (2003). Evaluation of physiological screening tests for breeding drought resistant triticale (x *Triticosecale* wittmack). *Acta Physiologiae Plantarum* 25: 29–37.

Guinn, G. (1976). Water deficit and ethylene evolution by young cotton bolls. *Plant Physiology* 57: 403–405.

Hake, K. and Silvertooth, J. (1990). High temperature effects on cotton. *Cotton Physiology Today Newsletter* 1: 1–4.

Hall, A.E. 2001. Heat stress and its impact. http://www.plantstress.com/Articles/heat_i/heat_i.htm, accessed 8th July 2019.

Hall, A.E. (2004). Breeding for adaptation to drought and heat in cowpea. *European Journal of Agronomy.* 21: 447–454.

Hall, A.E. and Allen, L.H. (1993). Designing cultivars for the climatic conditions of the next century. *International Journal of Crop Science*: 291–297.

Hasanuzzaman, M., Nahar, K., Alam, M.M. et al. (2013). Physiological, biochemical, and molecular mechanisms of heat stress tolerance in plants. *International Journal of Molecular Sciences* 14: 9643–9684.

Hatfield, J.L. and Burke, J.J. (1991). Energy exchange and leaf temperature behavior of three plant species. *Environmental and Experimental Botany* 31: 295–302.

Hatfield, J.L., Quisenberry, J.E., and Dilbeck, R.E. (1987). Use of canopy temperatures to identify water conservation in cotton germplasm. *Crop Science* 27: 269–273.

Havaux, I. (1993). Rapid photosynthetic adaptation to heat stress triggered in potato leaves by moderately elevated temperatures. *Plant, Cell & Environment* 16: 461–467.

Hofmann, W.C., Kittock, D.L., and Alemayehu, M. (1988). Planting seed density in relation to cotton emergence and yield. *Agronomy Journal* 80: 834–836.

Hong, S.W. and Vierling, E. (2000). Mutants of *Arabidopsis thaliana* defective in the acquisition of tolerance to high temperature stress. *Proceedings of the National Academy of Sciences* 97: 4392–4397.

Jamieson, P.D., Brooking, I.R., Porter, J.R., and Wilson, D.R. (1995). Prediction of leaf appearance in wheat: a question of temperature. *Field Crops Research* 41: 35–44.

Joshi, A.K. (1999). Genetic factors affecting abiotic stress tolerance in crop plants. In: *Handbook of Plant and Crop Stress* (ed. M. Pessarakli), 795–826. CRC Press.

Karademir, E., Karademir, Ç., Ekinci, R. et al. (2012). Screening cotton varieties (*Gossypium hirsutum* L.) for heat tolerance under field conditions. *African Journal of Agricultural Research* 7: 6335–6342.

Karim, M.A., Fracheboud, Y., and Stamp, P. (2000). Effect of high temperature on seedling growth and photosynthesis of tropical maize genotypes. *Journal of Agronomy and Crop Science* 184: 217–223.

Lather, B.P.S., Saini, M.L., and Punia, M.S. (2001). Hybrid cotton retrospect and prospects in Indian context. *National Journal of Plant Improvement* 3: 61–68.

Lin, J.J., Kuo, J., Ma, J. et al. (1996). Identification of molecular markers in soybean comparing RFLP, RAPD and AFLP DNA mapping techniques. *Plant Molecular Biology Reporter* 14: 156–169.

Liu, Z., Yuan, Y.L., Liu, S.Q. et al. (2006). Screening for high-temperature tolerant cotton cultivars by testing in vitro pollen germination, pollen tube growth and boll retention. *Journal of Integrative Plant Biology* 48: 706–714.

Lu, Z., Percy, R.G., Qualset, C.O., and Zeiger, E. (1998). Stomatal conductance predicts yields in irrigated Pima cotton and bread wheat grown at high temperatures. *Journal of Experimental Botany*: 453–460.

Maavimani, M., Jebaraj, S., Raveendran, M. et al. (2014). Cellular membrane thermostability is related to rice (*Oryza sativa* L.) yield under heat stress. *International Journal of Tropical Agriculture* 32: 201–208.

Maestri, E., Klueva, N., Perrotta, C. et al. (2002). Molecular genetics of heat tolerance and heat shock proteins in cereals. *Plant Molecular Biology* 48: 667–681.

Mahan, J.R., McMichael, B.L., and Wanjura, D.F. (1995). Methods for reducing the adverse effects of temperature stress on plants: a review. *Environmental and Experimental Botany* 35: 251–258.

Mahan, J.R., Young, A., Payton, P. et al. (2014). Effect of differential irrigation on accumulation of canopy temperature-based heat units in cotton. *Journal of Cotton Science* 18: 129–136.

Mao, L., Zhang, L., Zhao, X. et al. (2014). Crop growth, light utilization and yield of relay intercropped cotton as affected by plant density and a plant growth regulator. *Field Crops Research* 155: 67–76.

Marcum, K.B. (1998). Cell membrane thermostability and whole-plant heat tolerance of Kentucky bluegrass. *Crop Science* 38: 1214–1218.

Marsh, J.L., Walker, H., Theisen, H. et al. (2000). Expanded polyglutamine peptides alone are intrinsically cytotoxic and cause neurodegeneration in Drosophila. *Human Molecular Genetics* 9: 13–25.

Martineau, J.R., Specht, J.E., Williams, J.H., and Sullivan, C.Y. (1979). Temperature tolerance in soybeans: evaluation of a technique for assessing cellular membrane thermostability. *Crop Science* 19: 75–78.

Mauney, J.R., Kimball, B.A., Pinter, P.J. et al. (1994). Growth and yield of cotton in response to a free-air carbon dioxide enrichment (FACE) environment. *Agricultural and Forest Meteorology* 70: 49–67.

McMichael, B.L. and Burke, J.J. (1994). Metabolic activity of cotton roots in response to temperature. *Environmental and Experimental Botany* 34: 201–206.

Mickelbart, M.V., Hasegawa, P.M., and Bailey-Serres, J. (2015). Genetic mechanisms of abiotic stress tolerance that translate to crop yield stability. *Nature Reviews Genetics* 16: 237.

Mills, A.J., Cowling, R.M., Fey, M.V. et al. (2005). Effects of goat pastoralism on ecosystem carbon storage in semiarid thicket, Eastern Cape, South Africa. *Austral Ecology* 30: 797–804.

Monteith, J.L. (1995). A reinterpretation of stomatal responses to humidity. *Plant, Cell & Environment* 18: 357–364.

Morales, D., Rodriguez, P., Dell'Amico, J. et al. (2003). High-temperature preconditioning and thermal shock imposition affects water relations, gas exchange and root hydraulic conductivity in tomato. *Biologia Plantarum* 47: 203.

Nakamoto, H. and Hiyama, T. (1999). Heat-shock proteins and temperature stress. In: *Handbook of Plant and Crop Stress* (ed. M. Pessarakli), 399–416. New York: Marcel Dekker.

Newman, T.C., Ohme-Takagi, M., Taylor, C.B., and Green, P.J. (1993). DST sequences, highly conserved among plant SAUR genes, target reporter transcripts for rapid decay in tobacco. *Plant Cell* 5: 701–714.

Nielsen, K.F. (1974). Roots and root temperatures. *Proceedings of Plant Root Environment*: 293–333.

Niinemets, Ü. (2010). A review of light interception in plant stands from leaf to canopy in different plant functional types and in species with varying shade tolerance. *Ecological Research* 25: 693–714.

Paulsen, G.M. (1994). High temperature responses of crop plants. In: *Physiology and Determination of Crop Yield, (physiologyandde)* (eds. K.J. Boote, J.M. Bennett, T.R. Sinclair and G.M. Paulsen), 365–389.

Pettigrew, W.T. and Meredith, W.R. (1994). Leaf gas exchange parameters vary among cotton genotypes. *Crop Science* 34: 700–705.

Pettigrew, W.T. and Meredith, W.R. Jr. (2012). Genotypic variation in physiological strategies for attaining cotton lint yield production. *Journal of Cotton Science* 16: 179–189.

Pettigrew, W.T. and Turley, R.B. (1998). Variation in photosynthetic components among photosynthetically diverse cotton genotypes. *Photosynthesis Research* 56: 15–25.

Pettigrew, W.T., Heitholt, J.J., and Vaughn, K.C. (1993). Gas exchange differences and comparative anatomy among cotton leaf-type isolines. *Crop Science* 33: 1295–1299.

Poland, J.A. and Rife, T.W. (2012). Genotyping-by-sequencing for plant breeding and genetics. *The Plant Genome* 5: 92–102.

Queitsch, C., Hong, S.W., Vierling, E., and Lindquist, S. (2000). Heat shock protein 101 plays a crucial role in thermotolerance in *Arabidopsis*. *Plant Cell* 12: 479–492.

Quisenberry, J.E., McDonald, L.D., and McMichael, B.L. (1994). Responses of photosynthetic rates to genotypic differences in sink-to-source ratios in upland cotton (*Gossypium hirsutum* L.). *Environmental and Experimental Botany* 34: 245–252.

Radin, J.W., Kimball, B.A., Hendrix, D.L., and Mauney, J.R. (1987). Photosynthesis of cotton plants exposed to elevated levels of carbon dioxide in the field. *Photosynthesis Research* 12: 191–203.

Radin, J.W., Lu, Z., Percy, R.G., and Zeiger, E. (1994). Genetic variability for stomatal conductance in Pima cotton and its relation to improvements of heat adaptation. *Proceedings of the National Academy of Sciences* 91: 7217–7221.

Rahman, H.U. (2005). Genetic analysis of stomatal conductance in upland cotton (*Gossypium hirsutum* L.) under contrasting temperature regimes. *Journal of Agricultural Science* 143: 161–168.

Rahman, H.U., Malik, S.A., and Saleem, M. (2004). Heat tolerance of upland cotton during the fruiting stage evaluated using cellular membrane thermostability. *Field Crops Research* 85: 149–158.

Rashid, A., Stark, J.C., Tanveer, A., and Mustafa, T. (1999). Use of canopy temperature measurements as a screening tool for drought tolerance in spring wheat. *Journal of Agronomy and Crop Science* 182: 231–238.

Reddy, V.R., Reddy, K.R., and Baker, D.N. (1991). Temperature effect on growth and development of cotton during the fruiting period. *Agronomy Journal* 83: 211–217.

Rizhsky, L., Liang, H., and Mittler, R. (2002). The combined effect of drought stress and heat shock on gene expression in tobacco. *Plant Physiology* 130: 1143–1151.

Rizhsky, L., Liang, H., Shuman, J. et al. (2004). When defense pathways collide: the response of *Arabidopsis* to a combination of drought and heat stress. *Plant Physiology* 134: 1683–1696.

Sadras, V.O. (1995). Compensatory growth in cotton after loss of reproductive organs. *Field Crops Research* 40: 1–18.

Sawan, Z.M., Hanna, L.I., McCuistion, W.L., and Foote, R.J. (2009). Egyptian cotton (*Gossypium barbadense*) flower and boll production as affected by climatic factors and soil moisture status. *Theoretical and Applied Climatology* 99: 217–227.

Scharf, K.D., Siddique, M., and Vierling, E. (2001). The expanding family of *Arabidopsis thaliana* small heat stress proteins and a new family of proteins containing α-crystallin domains (Acd proteins). *Cell Stress & Chaperones* 6: 225–237.

Schöffl, F., Prandl, R., and Reindl, A. (1999). Molecular responses to heat stress. In: *Molecular Responses to Cold, Drought, Heat and Salt Stress in Higher Plants* (eds. K. Shinozaki and K. Yamaguchi-Shinozaki), 81–98. Austin, Texas: R.G. Landes Co.

Scott, J.W., Jones, J.B., Somodi, G.C., and Stall, R.E. (1995). Screening tomato accessions for resistance to *Xanthomonas campestris* pv. *vesicatoria*, race T3. *HortScience* 30: 579–581.

Scott, J.W., Stall, R.E., Jones, J.B., and Somodi, G.C. (1996). A single gene controls the hypersensitive response of Hawaii 7981 to race 3 (T3) of the bacterial spot pathogen. *Report of the Tomato Genetics Cooperative* 46: 23.

Singh, R.P., Prasad, P.V., Sunita, K. et al. (2007). Influence of high temperature and breeding for heat tolerance in cotton: a review. *Advances in Agronomy* 93: 313–385.

Song, G., Jiang, C., Ge, X. et al. (2014). Pollen thermotolerance of upland cotton related to anther structure and expression. *Agronomy Journal* 107: 1269–1279.

Sullivan, C.Y. (1972). Mechanisms of heat and drought resistance in grain sorghum and methods of measurement. In: *Sorghum in Seventies* (ed. N. Ganga Prasada Rao), 3–5. Oxford & IBH Pub. Co.

Taha, M.A., Malik, M.N.A., Chaudhry, F.I., and Makhdum, M.I. (1981). Heat-induced sterility in cotton sown during early April in West Punjab. *Experimental Agriculture* 17: 189–194.

Ulloa, M., Cantrell, R.G., Percy, R.G. et al. (2000). QTL analysis of stomatal conductance and relationship to lint yield in an interspecific cotton. *Journal of Cotton Science* 4: 10–18.

Wahid, A. and Shabbir, A. (2005). Induction of heat stress tolerance in barley seedlings by pre-sowing seed treatment with glycinebetaine. *Plant Growth Regulation* 46: 133–141.

Wahid, A., Gelani, S., Ashraf, M., and Foolad, M.R. (2007). Heat tolerance in plants: an overview. *Environmental and Experimental Botany* 61: 199–223.

Wanjura, D.F., Maas, S.J., Winslow, J.C., and Upchurch, D.R. (2004). Scanned and spot measured canopy temperatures of cotton and corn. *Computers and Electronics in Agriculture* 44: 33–48.

Yao, H., Zhang, Y., Yi, X. et al. (2016). Cotton responds to different plant population densities by adjusting specific leaf area to optimize canopy photosynthetic use efficiency of light and nitrogen. *Field Crops Research* 188: 10–16.

Zhang, W., Wang, Z., Yu, S. et al. (2004). Effects of planting density on canopy photosynthesis, canopy structure and yield formation of high-yield cotton in Xinjiang, China. *Acta Phytoecological Sinica* 28: 164–171.

2

Seed Priming as a Method to Generate Heat-stress Tolerance in Plants: A Minireview

Aditya Banerjee and Aryadeep Roychoudhury

Department of Biotechnology, St. Xavier's College (Autonomous), Kolkata, West Bengal, India

2.1 Introduction

Priming of seeds is often known as pre-treatment. This process accelerates some of the metabolic processes that are required for proper germination. Priming usually involves the use of solutions with high osmotic potential. Thus, in spite of promoting metabolic stimulation, the seeds cannot absorb enough water for the protrusion of radicles. The seeds are pushed to the lag phase of growth (Taylor et al. 1998). Seed priming markedly decreases the time required for the emergence of the seedlings after sowing. The advantage of seed pre-treatment compared to seedling treatment is lower resource requirement (Paul et al. 2017). A much lower amount of solution is required for treating the seeds compared to large volumes of solutions necessary for spraying the seedlings. Previous reports have shown that seed priming actually increased the yield of several crop species like wheat, barley, sorghum, maize, rice, pearl millet, chickpea, etc. (Banerjee and Roychoudhury 2018a).

The harmful effects of heat stress on crop physiology are major constraints on agricultural expansion in areas experiencing high temperatures. Several agronomic and genetic strategies have been undertaken for improving crop resilience to heat (Slafer and Savin 2018; Banerjee and Roychoudhury 2018b; Banerjee et al. 2016). Basic physiological abnormalities like altered membrane fluidity, reduced photosynthetic

Heat Stress Tolerance in Plants: Physiological, Molecular and Genetic Perspectives, First Edition.
Edited by Shabir Hussain Wani and Vinay Kumar.

efficiency, respiration and deteriorated water relations are usually observed in plant species which are susceptible to high temperatures (Banerjee and Roychoudhury 2016a,b; Banerjee et al. 2017; Roychoudhury and Banerjee 2016). Like any other abiotic stress, heat triggers uncontrolled production of the toxic reactive oxygen species (ROS) (Banerjee and Roychoudhury 2018c; Banerjee et al. 2018a). Accumulation of ROS beyond the threshold level causes inactivation of enzymes, damages the chromatin, and also promotes premature senescence (Banerjee and Roychoudhury 2018d,e). The effects of seed priming on heat stress are less investigated. This chapter presents a brief account of the ameliorative effects of various priming strategies in overcoming the physiological injuries caused by heat stress across plant species. Such approaches can be treated as economical, easy-to-use, farmer friendly as well as eco-friendly, without having to deal with any genetic modifications.

2.2 Mechanism of Heat Stress Injury in Plants

Extreme heat stress induces damage to cells and even necrosis, resulting in collapse of the cellular organization. Physiological processes like germination, growth, development, reproduction, and production are compromised (Lobell et al. 2011; Ahuja et al. 2010). Seedling establishment is reduced, owing to embryo damage resulting in low biomass accumulation (Cheng et al. 2009). The morphophysiological features like biomass development, root–water relations, phenology, and partitioning were largely affected in *Phaseolus vulgaris* seedlings exposed to heat stress (Koini et al. 2009). Tiller number is greatly reduced in heat stress. Extremely high temperatures delayed germination and triggered necrosis in *Triticum aestivum* plants. The enzyme structures were denatured and proteins were found to be aggregated. This resulted in aborted reproduction and compromised plant survival (Hasanuzzaman et al. 2013). C_3 plants are more affected by heat stress than C_4 plants are. High temperatures disorient the structural organization of thylakoids, resulting in altered granal stacking and granal swelling (Morales et al. 2003). The endogenous content of photosynthetic pigments along with the rates of CO_2 assimilation is also affected in heat susceptible cultivars (Marchand et al. 2005). The photochemistry in *Vitis vinifera* plants was greatly affected by heat stress as the average photosynthetic rate decreased by 60%, owing to an increase in temperature from 25 to 45 °C (Greer and Weedon 2012). Prolonged exposure to high temperatures reduced the content of soluble proteins, Rubisco-binding proteins (RBPs) and also inhibited the activities of crucial enzymes like sucrose phosphate synthase, ADP-glucose pyrophosphorylase, and invertase (Hasanuzzaman et al. 2013).

Among all the developmental stages, the phase of reproduction is most susceptible to variations in temperatures. Thus inhibited reproduction directly reduces crop yields. Heat stress aborts the development of floral buds and flowers. Fruits without any seeds have been observed in plant species exposed to heat stress (Maheswari et al. 2012; Foolad 2005). Impairment of pollen development, pollen tube growth, along with loss of ovule viability can be the reasons for sterility during heat stress (Cao et al. 2008). Rice seedlings grown under high temperatures produce excessively high quantities of ethylene which result in male sterility of the pollen grains. The sink strength in these plants is compromised by disturbed sugar-starch metabolism under the influence of excessive ethylene (Prasad et al. 2008). Heat stress negatively regulates the assimilatory capacity and alters membrane stability (Hay and Porter 2006). Reduction of radiation use efficiency (RUE) was decreased in the maize seedlings exposed to high temperatures ranging between 33 and 40 °C (Zhang et al. 2013). The characteristics of grain quality were greatly affected in barley plants grown under high temperature stress. The contents of carbohydrates like fructose, raffinose, and starch along with the concentrations of lipids and aluminum were greatly decreased (Vasseur et al. 2011).

High-temperature stress triggers the production of toxic ROS which promotes peroxidation of the lipids in cellular membrane systems (Banerjee and Roychoudhury 2018f; Banerjee et al. 2018b). ROS also causes large-scale damage to the photosynthetic machinery by degrading the chloroplast membrane and damaging the pigments (Banerjee and Roychoudhury 2018g). Owing to uncontrolled lipid peroxidation, the endogenous accumulation of malondialdehyde (MDA) increases (Hurkman et al. 2009). Such ROS-induced injuries resulted in premature senescence in cotton and wheat (Hasanuzzaman et al. 2013). ROS accumulation triggered membrane depolarization and activated the Ca^{2+}-induced RBOHD (respiratory burst oxidase homologue protein D) activity in the plasma membrane of rice (Qi et al. 2010).

2.3 Seed Priming Generating Heat-stress Tolerance

The adverse effects of heat stress can be ameliorated by using chemical agents to treat seeds before sowing them. Exogenous spraying of osmoprotectants, microelements, phytohormones, etc. at low concentrations has been useful for the generation of heat-stress tolerance in plants. However, the impacts of seed priming on heat-stress tolerance are less investigated. The plant hormone salicylic acid (SA) regulates multiple stress tolerance across a variety of species. Pre-treatment with SA ameliorated the injuries caused by heat stress in grapevine seedlings. The photosystem II (PSII) was protected resulting in accelerated recovery of the net photosynthetic rate (Pn) (Wang et al. 2010). Glycine-betaine (GB) is a compatible solute which confers osmoprotection against multiple stress-induced injuries (Roychoudhury and Banerjee 2016). Pre-sowing treatment of *Hordeum vulgare* seeds with 20 mM GB enhanced seed germination and improved fresh weight and dry weight. The rate of photosynthesis and leaf water potential increased, whereas the relative membrane permeability (RMP) decreased compared to the nontreated plants under heat stress (Wahid and Shabbir 2005). Seed priming with GB also reduced leakage of Ca^{2+}, K^+, and NO_3^- and created supra-optimal conditions for barley growth (Wahid and Shabbir 2005). Mahmood et al. (2016) report that seed priming with plant-growth-promoting rhizobacteria (biopriming) increased the rate and uniformity of germination along with improved harvest and yield in plants exposed to heat stress. This technology allows the bacteria to cling to seeds and acclimatize them to the sub-optimal conditions (Mahmood et al. 2016). Chen et al. (2012) report that seed osmopriming increased the accumulation of dehydrins and dehydrin-like proteins which promote tolerance to desiccation stress in *Spinacia oleracea*. Heat stress also causes desiccation of the tissues by inducing unwanted water loss via transpiration. Thus, such osmopriming strategies can be beneficial in promoting heat-stress tolerance in plants.

2.4 Conclusion

Heat stress is one of the most prevalent types of abiotic stress which directly induces plant desiccation. The plants experience severe osmotic stress and production of ROS, which damages the nucleic acid and protein structures. These results in uncontrolled necrosis, plant drooping, reduced yield, and ultimately plant mortality. Priming of seeds has been regarded as a useful technology in promoting stress tolerance against multiple abiotic stresses like salinity, drought, osmotic stress, etc. Seed pretreatment is economically cheap and requires low investments. In this technology, low amounts of solution need to be prepared in order to soak the seeds before sowing. Some reports on seed priming have been discussed which show that this technology can be useful to promote heat-stress tolerance across different plant species.

2.5 Future Perspectives

The generation of a heat-stress-tolerance trait by the method of seed priming is a rather naive, yet rapidly developing, field. Future perspectives would definitely involve the use of osmoprotectants like proline (Pro), polyamines (PAs), and other compatible solutes in seed priming for creating heat tolerance. These experiments would also facilitate the understanding of the roles of these biomolecules in the heat stress-responsive signaling pathway. Seed pretreatment with the universal stress hormone abscisic acid (ABA) might also yield heat tolerance, since ABA is known to participate in almost all abiotic stress-mediated responses (Roychoudhury and Banerjee 2017). Seed priming technology needs to be popularized as a cheap and useful method. This would benefit the farmers under actual field conditions and secure them from facing economic losses due to stress-induced compromised yield. Genome-wide transcriptomic analyses might also be undertaken to understand the involvement of various genes under the influence of different priming agents. This would effectively help in generating a signaling blueprint for a better understanding of heat-stress responses in plants.

Acknowledgments

Financial assistance from Council of Scientific and Industrial Research (CSIR), Government of India, through the research grant [38(1387)/14/EMR-II], Science and Engineering Research Board, Government of India through the grant [EMR/2016/004799] and Department of Higher Education, Science and Technology and Biotechnology, Government of West Bengal, through the grant [264(Sanc.)/ST/P/S&T/1G-80/2017] to Dr. Aryadeep Roychoudhury is gratefully acknowledged. Mr. Aditya Banerjee is thankful to University Grants Commission, Government of India for providing Junior Research Fellowship.

References

Ahuja, I., de Vos, R.C.H., Bones, A.M., and Hall, R.D. (2010). Plant molecular stress responses face climate change. *Trends in Plant Science* 15: 664–674.

Banerjee, A. and Roychoudhury, A. (2016a). Group II late embryogenesis abundant (LEA) proteins: structural and functional aspects in plant abiotic stress. *Plant Growth Regulation* 79: 1–17.

Banerjee, A. and Roychoudhury, A. (2016b). Plant responses to light stress: oxidative damages, photoprotection and role of phytohormones. In: *Plant Hormones Under Challenging Environmental Factors* (eds. G.J. Ahammed and J.-Q. Yu), 181–213. Dordrecht, The Netherlands: Springer.

Banerjee, A. and Roychoudhury, A. (2018a). Seed priming technology in the amelioration of salinity stress in plants.

In: *Advances in Seed Priming* (eds. A. Rakshit and H.B. Singh), 81–93. Singapore: Springer Nature.

Banerjee, A. and Roychoudhury, A. (2018b). Small heat shock proteins: structural assembly and functional responses against heat stress in plants. In: *Plant Metabolites and Regulation Under Abiotic Stress* (eds. P. Ahmad, M.A. Ahanger, V.P. Singh, et al.), 367–374. Academic Press/Elsevier.

Banerjee, A. and Roychoudhury, A. (2018c). Interactions of brassinosteroids with major phytohormones: antagonistic effects. *Journal of Plant Growth Regulation* https://doi.org/10.1007/s00344-018-9828-5.

Banerjee, A. and Roychoudhury, A. (2018d). The gymnastics of epigenomics in rice. *Plant Cell Reports* 37: 25–49.

Banerjee, A. and Roychoudhury, A. (2018e). Abiotic stress, generation of reactive oxygen species, and their consequences: an overview. In: *Revisiting the Role of Reactive Oxygen Species (ROS) in Plants: ROS Boon or Bane for Plants?* (eds. V.P. Singh, S. Singh, D. Tripathi, et al.), 23–50. Hoboken, NJ: Wiley.

Banerjee, A. and Roychoudhury, A. (2018f). Strigolactones: multi-level regulation of biosynthesis and diverse responses in plant abiotic stresses. *Acta Physiologiae Plantarum* 40: 86.

Banerjee, A. and Roychoudhury, A. (2018g). Regulation of photosynthesis under salinity and drought stress. In: *Environment and Photosynthesis: A Future Prospect* (eds. V.P. Singh, S. Singh, R. Singh and S.M. Prasad), 134–144. Studium Press (India) Pvt. Ltd.

Banerjee, A., Roychoudhury, A., and Krishnamoorthi, S. (2016). Emerging techniques to decipher microRNAs (miRNAs) and their regulatory role in conferring abiotic stress tolerance in plants. *Plant Biotechnology Reports* 10: 185–205.

Banerjee, A., Wani, S.H., and Roychoudhury, A. (2017). Epigenetic control of plant cold responses. *Frontiers in Plant Science* 8: 1643.

Banerjee, A., Tripathi, D.K., and Roychoudhury, A. (2018a). Hydrogen sulphide trapeze: environmental stress amelioration and phytohormone crosstalk. *Plant Physiology and Biochemistry* 132: 46–53.

Banerjee, A., Tripathi, D.K., and Roychoudhury, A. (2018b). The karrikin 'callisthenics': can compounds derived from smoke help in stress tolerance? *Physiologia Plantarum* https://doi.org/10.1111/ppl.12836.

Cao, Y.Y., Duan, H., Yang, L.N. et al. (2008). Effect of heat stress during meiosis on grain yield of rice cultivars differing in heat tolerance and its physiological mechanism. *Acta Agronomica Sinica* 34: 2134–2142.

Chen, K., Fessehaie, A., and Arora, R. (2012). Dehydrin metabolism is altered during seed osmopriming and subsequent germination under chilling and desiccation in *Spinacia oleracea* L. cv. Bloomsdale: possible role in stress tolerance. *Plant Science* 183: 27–36.

Cheng, L., Zou, Y., Ding, S. et al. (2009). Polyamine accumulation in transgenic tomato enhances the tolerance to high temperature stress. *Journal of Integrative Plant Biology* 51: 489–499.

Foolad, M.R. (2005). Breeding for abiotic stress tolerances in tomato. In: *Abiotic Stresses: Plant Resistance Through Breeding and Molecular Approaches* (eds. M. Ashraf and P.J.C. Harris), 613–684. New York: The Haworth Press Inc.

Greer, D.H. and Weedon, M.M. (2012). Modelling photosynthetic responses to temperature of grapevine (*Vitis vinifera* cv. Semillon) leaves on vines grown in a hot climate. *Plant, Cell & Environment* 35: 1050–1064.

Hasanuzzaman, M., Nahar, K., Alam, M.M. et al. (2013). Physiological, biochemical, and molecular mechanisms of heat stress tolerance in plants. *International Journal of Molecular Sciences* 14: 9643–9684.

Hay, R.K.M. and Porter, J.R. (2006). *The Physiology of Crop Yield*. Oxford: Blackwell Publishing Ltd.

Hurkman, W.J., Vensel, W.H., Tanaka, C.K. et al. (2009). Effect of high temperature on albumin and globulin accumulation in the endosperm proteome of the developing wheat grain. *Journal of Cereal Science* 49: 12–23.

Koini, M.A., Alvey, L., Allen, T. et al. (2009). High temperature-mediated adaptations in plant architecture require the bHLH transcription factor PIF4. *Current Biology* 19: 408–413.

Lobell, D.B., Schlenker, W., and Costa-Roberts, J. (2011). Climate trends and global crop production since 1980. *Science* 333: 616–620.

Maheswari, M., Yadav, S.K., Shanker, A.K. et al. (2012). Overview of plant stresses: mechanisms, adaptations and research pursuit. In: *Crop Stress and Its Management: Perspectives and Strategies* (eds. B. Venkateswarlu, A.K. Shanker, C. Shanker and M. Maheswari), 1–18. Dordrecht, The Netherlands: Springer.

Mahmood, A., Turgay, O.C., Farooq, M., and Hayat, R. (2016). Seed biopriming with plant growth promoting rhizobacteria: a review. *FEMS Microbiology Ecology* 92.

Marchand, F.L., Mertens, S., Kockelbergh, F. et al. (2005). Performance of High Arctic tundra plants improved during but deteriorated after exposure to a simulated extreme temperature event. *Global Change Biology* 11: 2078–2089.

Morales, D., Rodriguez, P., Dellamico, J. et al. (2003). High-temperature preconditioning and thermal shock imposition affects water relations, gas exchange and root hydraulic conductivity in tomato. *Biologia Plantarum* 47: 203–208.

Paul, S., Roychoudhury, A., Banerjee, A. et al. (2017). Seed pre-treatment with spermidine alleviates oxidative damages to different extent in the salt (NaCl)-stressed seedlings of three Indica rice cultivars with contrasting level of salt tolerance. *Plant Gene* 11: 112–123.

Prasad, P.V.V., Staggenborg, S.A., and Ristic, Z. (2008). Impacts of drought and/or heat stress on physiological, developmental, growth, and yield processes of crop plants. In: *Response of Crops to Limited Water: Understanding and Modeling Water Stress Effects on Plant Growth Processes*, Advances in Agricultural Systems Modeling Series, vol. 1 (eds. L.H. Ahuja and S.A. Saseendran), 301–355. Madison, WI: ASA-CSSA.

Qi, Y., Wang, H., Zou, Y. et al. (2010). Over-expression of mitochondrial heat shock protein 70 suppresses programmed cell death in rice. *FEBS Letters* 585: 231–239.

Roychoudhury, A. and Banerjee, A. (2016). Endogenous glycine betaine accumulation mediates abiotic stress tolerance in plants. *Tropical Plant Research* 3: 105–111.

Roychoudhury, A. and Banerjee, A. (2017). Abscisic acid signaling and involvement of mitogen activated protein kinases and calcium-dependent protein kinases during plant abiotic stress. In: *Mechanism of Plant Hormone Signaling Under Stress*, vol. 1 (ed. G. Pandey), 197–241. Hoboken, NJ: Wiley.

Slafer, G.A. and Savin, R. (2018). Can N management affect the magnitude of yield loss due to heat waves in wheat and maize? *Current Opinion in Plant Biology* https://doi.org/10.1016/j.pbi.2018.07.009.

Taylor, A.G., Allen, P.S., Bennett, M.A. et al. (1998). Seed enhancements. *Seed Science Research* 8: 245–256.

Vasseur, F., Pantin, F., and Vile, D. (2011). Changes in light intensity reveal a major role for carbon balance in *Arabidopsis* responses to high temperature. *Plant, Cell & Environment* 34: 1563–1576.

Wahid, A. and Shabbir, A. (2005). Induction of heat stress tolerance in barley seedlings by pre-sowing seed treatment with glycinebetaine. *Plant Growth Regulation* 46: 133–141.

Wang, L.J., Fan, L., Loescher, W. et al. (2010). Salicylic acid alleviates decreases in photosynthesis under heat stress and accelerates recovery in grapevine leaves. *BMC Plant Biology* 10: 34.

Zhang, X., Cai, J., Wollenweber, B. et al. (2013). Multiple heat and drought events affect grain yield and accumulations of high molecular weight glutenin subunits and glutenin macropolymers in wheat. *Journal of Cereal Science* 57: 134–140.

3

How Effective Are Stress-associated Proteins in Augmenting Thermotolerance?

Inès Karmous and Sandeep Kumar Verma

Biotechnology Laboratory, Department of Biology, Bolu Abant Izzet Baysal University, Bolu, Turkey

3.1 Introduction

In order to ensure their growth and development, plants need specific optimal temperatures. The elevation or decrease of temperatures has been currently recognized as among the most drastic problems in our environment, owing to ongoing climatic change. In plants, metabolic processes require an optimal temperature of 30 °C, and temperatures below 30 °C and above 40 °C are commonly considered stress condition. Generally, heat stress is defined as the increase in temperature by 10–15 °C compared with optimal temperature. Such elevation can occur either gradually (heat stress) or suddenly (heat shock), and may result in reversible or irreversible alterations of the physiological, metabolic, and cellular functions (Wahid et al. 2007; Nguyen et al. 2013).

Tolerance consists of the development, activation, or modification of physiological mechanisms that allow the plant to either resist stress without the onset of injuries or repair the damage. Thus, the level of resistance of a plant to a given stress will depend on its ability to activate tolerance mechanisms. However, most of the resistance processes described in plants are referred to as physiological and biochemical mechanisms that allow them to tolerate the effects of temperature changes. Eventually, thermotolerance would be conferred through the

Heat Stress Tolerance in Plants: Physiological, Molecular and Genetic Perspectives, First Edition.
Edited by Shabir Hussain Wani and Vinay Kumar.

expression of specific genes encoding proteins, enzymes, and molecular antioxidants (Howe and Schilmiller 2002), triggering specific metabolic and physiological reactions, or even encoding some adaptative morphological and physiological characters. In this context, the present chapter aims to review the different types of stress-related proteins in response to heat stress, and discuss their roles in augmenting potential for adaptation in response to thermal stress.

3.1.1 Heat Shock Proteins (HSPs)

Among the stress proteins, HSPs are either activated/stimulated or induced under heat stress. Indeed, certain preexisting proteins under normal condition increase by levels of expression (overexpression), while other proteins are synthesized *de novo* after exposure to stress (Falfushynska et al. 2016). HSPs are considered among the thermo-induced proteins, belonging to molecular chaperones (Boston et al. 1996; Collada et al. 1997). Induction of HSP expression or synthesis was shown to be a rapid cellular response, and stimulated at temperatures exceeding 8–10 °C of optimal temperature. It may occur a few seconds after the temperature increase, and the maximum level of transcripts may be observed within one to two hours of exposure, when it begins to decline (Larkindale et al. 2005). This involves the generation of polypeptides of two different molecular weights with sizes ranging from 15 to 30 kDa, localized in different subcellular compartments: cytoplasm, nucleus, endoplasmic reticulum (ER), mitochondria, and chloroplast (Van Montfort et al. 2001). In fact, HSPs can be assigned to five families of proteins identified in bacteria, yeast, plants, and animals, based on sequence similarity and differences in molecular weight: HSP100, HSP90, HSP70, HSP60, and small HSPs. In plants, only the HSP100 family has been shown to be essential for the acquisition of thermotolerance by genetic analysis (Hong and Vierling 2001).

High temperatures are known to affect proteins' stability and cause their dysfunction. Hence, maintaining proteins in their functional conformations and preventing the aggregation of non-native proteins are therefore important for cell survival under stress. Heat-shock proteins (HSPs)/chaperones are responsible for protein folding, assembly, translocation, and degradation in many normal cellular processes. They can play a crucial role in protecting physiological processes against stress, and assist in various processes involved in the maintenance of heat responses (Shinozaki et al. 2015). Among HSP roles, their involvement in re-establishing normal protein conformation and refolding helps to protect cellular homeostasis. They favor the stabilization of proteins structures, by maintaining the functional conformations of proteins and of certain polypeptides that would interact/associate with other functional components or of the transport systems or other associated polypeptides. They are also involved in the elimination of aggregated and denatured proteins, and nonfunctional or potentially dangerous polypeptides, preventing irreversible protein aggregation (Van Montfort et al. 2001). HSP-bound proteins are subsequently refolded by ATP-dependent chaperones of the DnaK/HSP70 system (Lee and Vierling 2000). They are involved in preserving membrane integrity during thermal fluctuations (Tsvetkova et al. 2002). HSPs also play important roles in transmembrane transports and the production of multimeric protein complexes, such as ribulose biphosphate carboxylase/oxygenase in plants. For example, chloroplast small HSP (smHSP) are shown to exhibit a protective role on photosynthetic electron transport, during heat, oxidative, and photoinhibition stresses (Shakeel et al. 2012). A recent study showed that HSP21 expression in the wild genotype of *Arabidopsis thaliana* allowed recovery and acquisition of thermotolerance (Sedaghatmehr et al. 2016).

Hence, the expression of HSPs is considered a strategy of adaptation to heat stress, which plays a role in conferring thermotolerance in many plant species (Vierling 1991; Wahid et al. 2007). HSP gene expression is regulated mainly at the transcriptional level. Upon heat stress, heat shock transcription factors (HSFs) in cytoplasm become active through oligomerization and recompartmentation to the nucleus (Nover and Scharf 1997), thus modulating transcription during long-term heat stress response (Nover et al. 2001). Studies conducted on wheat seeds report that high temperatures induce the synthesis of HSPs of molecular weights 111, 90, 70, 45, 32, 24, and 8 KDa. Similarly, other studies have shown that plants exposed to heat stress induced the expression (massive transcription and translation) of HSPs (Boston et al. 1996). Lee et al. (2007) found that HSPs and antioxidant enzymes were up-regulated under heat stress in rice leaves. Similarly, Li et al. (2013) identify stress-related proteins possibly involved in mechanisms of stress adaptation and investigated their expression (up-regulated or down-regulated) in alfalfa. Proteins up-regulated were reported to be related to energy (photosynthesis, ATP synthase), signal transduction, intracellular traffic, and disease/defense, ubiquitin, alcohol dehydrogenase, methyl binding domain protein, and HSP70. On the other hand, the proteomic analysis (two-dimensional electrophoresis with mass spectroscopy) has revealed the types of proteins and pathways conferring thermotolerance (Zou et al. 2011; Huang et al. 2012). Other studies demonstrate the involvement of HSPs in cold climates in trees (Lubaretz and Zur Nieden 2002; López-Matas et al. 2004). Among proteins induced or up-regulated in plants by low temperatures, there have been reported those of high molecular weight HSPs (Sung et al. 2001) and also small heat shock proteins (sHSPs) (Ukaji et al. 1999). So far, woody species in which cold-specific accumulation of HSPs has been observed include *Prunus persica* (Wisniewski et al. 1996), *Castanea sativa* (López-Matas et al. 2004). *Morus bombycis* (Ukaji et al. 1999), and *Acer platanoides* and *Sambucus nigra* (Lubaretz and Zur Nieden 2002). Nonetheless, stress conditions other than elevated temperatures can also lead to Hsp induction. This includes osmotic and salt stress, stress from low oxygen, toxic compounds, dinitrophenol, UV radiation, and treatment with plant hormones such as abscisic acid or ethylene (Boston et al. 1996; Iba 2002).

3.1.2 Proline

Proline is among those osmolytes that act as buffers to redox potential and prevent oxidative damage (Anwar Hossain et al. 2014). It plays a role in metabolic signalization, and controls different functions such as development, stress resistance/tolerance, and regulation of the redox balance under stress conditions (Wahid and Close 2007; Rasheed et al. 2011; Hasanuzzaman et al. 2013). It is involved in the protection of cellular structures, stability and integrity of proteins, and increase of activity of enzymes (Lee and Suh 2013). The increase of the level of proline is generally considered an osmoprotector, and thus as a criterion of adaptation to heat stress (Dhyani et al. 2013).

3.1.3 Dehydrins (DHNs)

Dehydrins (DHNs) are proteins that have been shown to play a protective role during cellular dehydration (Close 1997; Campbell and Close 1997). Studies suggest that DHNs can stabilize membranes and rescue hydrolytic enzyme function (Danyluk et al. 1998; Rinne et al. 1999). Authors have shown the overexpression of dehydrin genes increases freeze tolerance in diverse species (Karlson et al. 2003; Puhakainen et al. 2004), with differential responses of these different genes of DHN family to cold (Welling et al. 2002; Karlson et al. 2003; Welling et al. 2004).

3.1.4 Role of Metabolic Proteins in Thermotolerance

Under thermal stress, overproduction of reactive oxygen species (ROS) is one of the main metabolic damages caused by exposure to heat (Mittler et al. 2011; Suzuki et al. 2012, 2013). In order to cope with such stress conditions, plants develop different strategies to activate the potential for rapid evolution of thermal tolerance, notably defense mechanisms to prevent ROS damage. This capacity is dependent on the heritable genetic variation in thermal sensitivity within plant populations. Indeed, ROS, by accumulating in cell compartments and components, mainly in the chloroplasts, mitochondria, and peroxisomes, can cause oxidative damage, affecting structural and functional proteins, including enzymes. Besides, any damage within the structure of the enzymatic proteins leads to their denaturation, oxidation of main amino acids, and changes in their spatial conformation, which causes their deactivation or inactivation (Źróbek-Sokolnik 2012; Taiz and Zeiger 2015). Hence, a protection of the functional structure of enzymatic proteins is of important interest, in order to control ROS production, and maintain redox homeostasis and metabolic activities (De Pinto et al. 2015; Zou et al. 2017).

Therefore, the mechanism to overcome heat tolerance involves changes in proteins related to the protection of proteins, protein and enzyme biosynthesis/degradation, and redox homeostasis (Yaneva et al. 2002; Gao et al. 2013; Zou et al. 2011). In addition, antioxidant enzymes play a key role in ameliorating the antioxidative capacity of cells. Hence, the overexpression of these metabolic proteins, notably the antioxidant and redox enzymes, as well as the activation of their activities, may contribute to counteracting oxidative damage, by ROS scavenging. For example, the role of catalase, dismutase superoxide, peroxidases, and glutathione-S-transferases (GSTs) (Board and Menon 2013; Zhang et al. 2015; Ali et al. 2016). In other studies, total antioxidant capacity (T-AOC) has been shown to adapt to deal with oxidative stress and free radical formation and was consistent with the results reported by Zhang et al. (2015, 2014). Likewise, many reports suggest that maintaining homeostasis and steady-state physiology of the plant enable it to withstand the rise or decrease of temperature (De Pinto et al. 2015; Shinozaki et al. 2015).

Regulation of nitrogen and sugar metabolisms as well as energy production to plants (photosynthesis and carbon metabolism) are important factors in heat tolerance. Basically, enzymes involved in such metabolisms undergo down- or up-regulation. For example, 5-aminolevulinate synthase acts in the overall capacity of chlorophyll formation in heat-stressed wheat seedlings (Mohanty et al. 2006). Additionally, Li et al. (2013) identified proteins involved in different metabolic processes, in particular tricarboxylic acid (TCA) cycle, glycolysis, such as photosystem I reaction center subunit II, Rubisco activase, ribulose-1,5-bisphosphate carboxylase/oxygenase (Rubisco), chlorophyll a/b binding protein (LHCII), oxygen-evolving enhancer protein (OEE), and plastocyanin. These authors showed the up-regulation of glyceraldehyde-3-phosphate dehydrogenase, fructose bisphosphate aldolase, and Rubisco enzyme under heat shock, which led to increased abundance and activity of these enzymes. This may maintain the CO_2 assimilation rate and stability of photosystems and enhance photosynthesis. On the other hand, plastocyanin is also an essential protein that donates electrons to PSI. Its activity was evidenced by the level of its expression (up-regulation/down-regulation), which controls the electron transfer to photosystems, the structure, and the total rate of energy transduction in a photosynthetic membrane.

It is important to point that the level of expression of proteins varies within species and with types of environmental stresses. The up-regulation of some proteins, particularly enzymatic proteins, involved in the photosynthesis process and energy metabolism provides the energy required to tolerate heat stress. For examples, (i) the expression of OEE proteins were shown to be correlated with abiotic stresses such as cold stress (Rinalducci et al. 2011) and heat stress (Zhang et al. 2013), (ii) the overexpression of enzymes involved in TCA cycle, including citrate synthase, oxoglutarate dehydrogenase, fumarate hydratase, and malate dehydrogenase; in glycolysis, including FBA (fructose bisphosphate aldolase), GADPH (glyceraldehyde-3-phosphate dehydrogenase), and enolase; and other proteins, including nucleoside diphosphate kinase, cupin-domain-containing protein, NAD (nicotinamide adenine dinucleotide) dependent dehydrogenase, ATP (adenosine triphosphate) synthase b subunit that play key roles in maintaining cellular levels of different energy forms responsible for enhancing multiple stress tolerance (Moon et al. 2003). Likewise, cupin-domain-containing protein has biochemical activities associated with the cell wall. This protein protects other proteins against thermal denaturation, thus it can exhibit the ability to protect other proteins against thermal denaturation (Pelah et al. 1995). Other studies revealed the down-regulation of other key enzymes of the energy metabolism, such as fumarate hydratase, malate dehydrogenase, and oxoglutarate dehydrogenase, suggesting the inhibition of carbohydrate metabolism that may further affect energy-dependent processes involved in heat stress resistance.

On the other hand, the regulation of the heat stress response can also be carried out via the up-regulation of some proteases and proteins involved in the ubiquitin–proteasome pathway. For examples, kinase peptidyl-prolyl cis-trans isomerase and protein disulfide isomerase, which catalyze protein folding, peptidyl-prolyl cis-trans isomerase which accelerates the folding of proteins (Pemberton 2006) and protein disulfide isomerase that is involved in the post-translational modification disulfide exchange (Hatahet and Ruddock 2009). In studies investigating the basic molecular mechanisms of cold acclimation in plants, researchers have reported the induction of protective proteins and the change in carbohydrate metabolism (Kozlowski and

Pallardy 2002; Arora et al. 2003). The role of the increase in the activity of the enzyme sucrose-phosphate synthase, key enzymes of sucrose synthesis, in sugar accumulation, and acquisition of cold tolerance have also been reported.

3.2 Conclusion

Different lines of evidence indicate heat stress has a complex influence on cell function and homeostasis, suggesting that thermotolerance involves many processes. Some of them may probably involve proteins with regulatory metabolic function or be engaged in signal transduction pathways, such as protein kinases and hormone biosynthetic enzymes. However, more research is required to understand the molecular response and the modulation mechanism to heat tolerance in plants, notably in identifying genes and proteins so far that contribute significantly to the development of thermotolerance. Genes encoding these stress-related proteins require more investigation, using the tools of forward and reverse genetics approaches, such as genetic engineering, or novel alleles can be sought through germplasm screening or mutagenesis.

References

Ali, A., Rashid, M.A., Huang, Q.Y. et al. (2016). Response of antioxidant enzymes in *Mythimna separata* (Lepidoptera: Noctuidae) exposed to thermal stress. *Bulletin of Entomological Research* 107 (3) https://doi.org/10.1017/S0007485316001000.

Anwar Hossain, M., Hoque, M.A., Burritt, D.J., and Fujita, M. (2014). Proline protects plants against abiotic oxidative stress. In: *Oxidative Damage to Plants* (ed. P. Ahmad), 477–522. San Diego: Elsevier.

Arora, R., Rowland, L.J., and Tanino, K. (2003). Induction and release of bud dormancy in woody perennials: a science comes of age. *HortScience* 38: 911–921.

Board, P.G. and Menon, D. (2013). Glutathione transferases, regulators of cellular metabolism and physiology. *Biochimica et Biophysica Acta* 1830: 3267–3288.

Boston, R.S., Viitanen, P.V., and Vierling, E. (1996). Molecular chaperones and protein folding in plants. *Plant Molecular Biology* 32: 191–222.

Campbell, S.A. and Close, T.J. (1997). Dehydrins: genes, proteins, and associations with phenotypic traits. *The New Phytologist* 137: 61–74.

Close, T.J. (1997). Dehydrins: a commonalty in the response of plants to dehydration and low temperature. *Physiologia Plantarum* 100: 291–296.

Collada, C., Gomez, L., Casado, R., and Aragoncillo, C. (1997). Purification and in vitro chaperone activity of a class I small heat-shock protein abundant in recalcitrant chestnut seeds. *Plant Physiology* 115: 71–77.

Danyluk, J., Perron, A., Houde, M. et al. (1998). Accumulation of an acidic dehydrin in the vicinity of the plasma membrane during cold acclimation of wheat. *The Plant Cell* 10: 623–638.

De Pinto, M.C., Locato, V., Paradiso, A., and De Gara, L. (2015). Role of redox homeostasis in thermo- tolerance under a climate change scenario. *Annals of Botany* 116 (4): 487–496. https://doi.org/10.1093/aob/mcv071.

Dhyani, K., Ansari, M.W., Rao, Y.R. et al. (2013). Comparative physiological response of wheat genotypes under terminal heat stress. *Plant Signaling & Behavior* 8: e24564.

Falfushynska, H.I., Phan, T., and Sokolova, I.M. (2016). Long-term acclimation to different thermal regimes affects molecular responses to heat stress in a freshwater clam *Corbicula fluminea*. *Scientific Reports* 6: 39476. https://doi.org/10.1038/srep39476.

Gao, H.B., Chu, Y.J., and Xue, H.W. (2013). Phosphatidic acid (PA) Binds PP2AA1 to regulate PP2A activity and PIN1 polar localization. *Molecular Plant* 6 (5): 1692–1702. https://doi.org/10.1093/mp/sst076.

Hasanuzzaman, M., Nahar, K., Alam, M. et al. (2013). Physiological, biochemical, and molecular mechanisms of heat stress tolerance in plants. *International Journal of Molecular Sciences* 14 (5): 9643–9684. https://doi.org/10.3390/ijms14059643.

Hatahet, F. and Ruddock, L.W. (2009). Protein disulfide isomerase: a critical evaluation of its function in disulfide bond formation. *Antioxidants & Redox Signaling* 11: 2807–2850.

Hong, S.W. and Vierling, E. (2001). Hsp101 is necessary for heat tolerance but dispensable for development and germination in the absence of stress. *The Plant Journal* 27: 25–35.

Howe, G.A. and Schilmiller, A.L. (2002). Oxylipin metabolism in response to stress. *Current Opinion in Plant Biology* 5: 230–236.

Huang, B., Rachmilevitch, S., and Xu, J. (2012). Root carbon and protein metabolism associated with heat tolerance. *Journal of Experimental Botany* 63: 3455–3465.

Iba, K. (2002). Acclimative response to temperature stress in higher plants: approaches of gene engineering for temperature tolerance. *Annual Review of Plant Biology* 53: 225–245.

Karlson, D.T., Zeng, Y., Stirm, V.E. et al. (2003). Photoperiodic regulation of a 24-kD dehydrin-like protein in red-osier dogwood (*Cornus sericea* L.) in relation to freeze-tolerance. *Plant and Cell Physiology* 44: 25–34.

Kozlowski, T. and Pallardy, S. (2002). Acclimation and adaptive responses of woody plants to environmental stresses. *The Botanical Review* 68: 270–334.

Larkindale, J., Mishkind, M., and Vierling, E. (2005). Plant responses to high temperature. In: *Plant Abiotic Stress* (eds. M. Jenks and P.M. Hasegawa), 100–134. Oxford: Blackwell.

Lee, S.B. and Suh, M.C. (2013). Recent advances in cuticular wax biosynthesis and its regulation in *Arabidopsis*. *Molecular Plant* 6: 246–249.

Lee, G.J. and Vierling, E. (2000). A small heat-shock protein cooperates with heat shock protein 70 systems to reactivate a heat-denatured protein. *Plant Physiology* 122: 189–198.

Lee, D.G., Ahsan, N., Lee, S.H. et al. (2007). A proteomic approach in analyzing heat-responsive proteins in rice leaves. *Proteomics* 7: 3369–3383.

Li, W., Wei, Z., Qiao, Z. et al. (2013). Proteomics analysis of alfalfa response to heat stress. *PLoS One* 8: e82725.

López-Matas, M.A., Nuñez, P., Soto, A. et al. (2004). Protein cryoprotective activity of a cytosolic small heat shock protein that accumulates constitutively in chestnut stems and is up-regulated by low and high temperatures. *Plant Physiology* 134: 1708–1717.

Lubaretz, O. and Zur Nieden, U. (2002). Accumulation of plant small heat-stress proteins in storage organs. *Planta* 215: 220–228.

Mittler, R., Vanderauwera, S., Suzuki, N. et al. (2011). ROS signaling: the new wave? *Trends in Plant Science* 16 (6): 300–309. https://doi.org/10.1016/j.tplants.2011.03.007.

Mohanty, S., Grimm, B., and Tripathy, B.C. (2006). Light and dark modulation of chlorophyll biosynthetic genes in response to temperature. *Planta* 224: 692–699.

Moon, H., Lee, B., Choi, G. et al. (2003). NDP kinase 2 interacts with two oxidative stress-activated MAPKs to regulate cellular redox state and enhances multiple stress tolerance in transgenic plants. *Proceedings of the National Academy of Sciences of the United States of America* 100: 358–363.

Nguyen, T.M., Bressac, C., and Chevrier, C. (2013). Heat stress affects male reproduction in a parasitoid wasp. *Journal of Insect Physiology* 59: 248–254.

Nover, L. and Scharf, K.D. (1997). Heat stress proteins and transcription factors. *Cellular and Molecular Life Sciences* 53: 80–103.

Nover, L., Bhartı, K., Dörıng, P. et al. (2001). *Arabidopsis* and the Hsf world: how many heat stress transcription factors do we need? *Cell Stress & Chaperones* 6: 177–189.

Pelah, D., Shoseyov, O., and Altman, A. (1995). Characterization of BspA, a major boiling-stable, water-stress-responsive protein in aspen (*Populus tremula*). *Tree Physiology* 15: 673–678.

Pemberton, T.J. (2006). Identification and comparative analysis of sixteen fungal peptidyl-prolyl cis/trans isomerase repertoires. *BMC Genomics* 7: 244.

Puhakainen, T., Hess, M.W., Mäkelä, P. et al. (2004). Overexpression of multiple dehydrin genes enhances tolerance to freezing stress in *Arabidopsis*. *Plant Molecular Biology* 54: 743–753.

Rasheed, R., Wahid, A., Farooq, M. et al. (2011). Role of proline and glycinebetaine pretreatments in improving heat tolerance of sprouting sugarcane (*Saccharum* sp.) buds. *Plant Growth Regulation* 65 (1): 35–45. https://doi.org/10.1007/s10725-011-9572-3.

Rinalducci, S., Egidi, M.G., Karimzadeh, G. et al. (2011). Proteomic analysis of a spring wheat cultivar in response to prolonged cold stress. *Electrophoresis* 32: 1807–1818.

Rinne, P.L., Kaikuranta, P.L., and van der Plas, L.H. (1999). Dehydrins in cold-acclimated apices of birch (*Betula pubescens*

Ehrh.): production, localization and potential role in rescuing enzyme function during dehydration. *Planta* 209: 377–388.

Sedaghatmehr, M., Mueller-Roeber, B., and Balazadeh, S. (2016). The plastid metalloprotease FtsH6 and small heat shock protein HSP21 jointly regulate thermomemory in *Arabidopsis. Nature Communications* 7 (1): 12439. https://doi.org/10.1038/ncomms12439.

Shakeel, S.N., Haq, N.U., Heckathorn, S., and Luthe, D. (2012). Analysis of gene sequences indicates that quantity not quality of chloroplast small HSPs improves thermotolerance in C4 and CAM plants. *Plant Cell Reports* 31: 1943–1957.

Shinozaki, K., Uemura, M., Bailey-Serres, J. et al. (2015). Responses to abiotic stress. In: *Biochemistry and Molecular Biology of Plants* (eds. B.B. Buchanan, W. Gruissem, K. Vickers and R.L. Jones). Hoboken, NJ: Wiley.

Sung, D.Y., Vıerlıng, E., and Guy, C.L. (2001). Comprehensive expression profile analysis of the Arabidopsis Hsp70 gene family. *Plant Physiology* 126: 789–800.

Suzuki, N., Koussevitzky, S., Mittler, R., and Miller, G. (2012). ROS and redox signalling in the response of plants to abiotic stress. *Plant, Cell and Environment* 35 (2): 259–270. https://doi.org/10.1111/j.1365-3040.2011.02336.x.

Suzuki, N., Miller, G., Salazar, C. et al. (2013). Temporal-spatial interaction between reactive oxygen species and abscisic acid regulates rapid systemic acclimation in plants. *The Plant Cell* 25 (9): 3553–3569. https://doi.org/10.1105/tpc.113.114595.

Taiz, L. and Zeiger, E. (2015). *Plant Physiology and Development*, 6e. Sinauer Associates, Inc.

Tsvetkova, N.M., Horväth, I., Török, Z. et al. (2002). Small heat shock proteins regulate membrane lipid polymorphism. *Proceedings of the National Academy of Sciences of the United States of America* 99: 13504–13509.

Ukaji, N., Kuwabara, C., Takezawa, D. et al. (1999). Accumulation of small heat-shock protein homologs in the endoplasmic reticulum of cortical parenchyma cells in mulberry in association with seasonal cold acclimation. *Plant Physiology* 120: 481–489.

Van Montfort, R., Slıngsby, C., and Vıerlıng, E. (2001). Structure and function of the small heat shock protein/alpha-crystallin family of molecular chaperones. *Advances in Protein Chemistry* 59: 105–156.

Vierling, E. (1991). The roles of heat shock proteins in plants. *Annual Review of Plant Physiology and Plant Molecular Biology* 42 (1): 579–620. https://doi.org/10.1146/annurev.pp.42.060191.003051.

Wahid, A. and Close, T.J. (2007). Expression of dehydrins under heat stress and their relationship with water relations of sugarcane leaves. *Biologia Plantarum* 51 (1): 104–109. https://doi.org/10.1007/s10535-007-0021-0.

Wahid, A., Gelani, S., Ashraf, M., and Foolad, M.R. (2007). Heat tolerance in plants: an overview. *Environmental and Experimental Botany* 61: 199–223.

Welling, A., Moritz, T., Palva, E.T., and Junttila, O. (2002). Independent activation of cold acclimation by low temperature and short photoperiod in hybrid aspen. *Plant Physiology* 129: 1633–1641.

Welling, A., Rinne, P., Viherä-Aarnio, A. et al. (2004). Photoperiod and temperature differentially regulate the expression of two dehydrin genes during overwintering of birch (*Betula pubescens* Ehrh.). *Journal of Experimental Botany* 55: 507–516.

Wisniewski, M., Close, T.J., Artlıp, T., and Arora, R. (1996). Seasonal patterns of dehydrins and 70-kDa heat-shock proteins in bark tissues of eight species of woody plants. *Physiologia Plantarum* 96: 496–505.

Yaneva, I.A., Hoffmann, G.W., and Tischner, R. (2002). Nitrate reductase from winter wheat leaves is activated at low temperature via protein dephosphorylation. *Physiologia Plantarum* 114: 65–72. https://doi.org/10.1034/j.1399-3054.2002.1140110.x.

Zhang, Y., Xu, L., Zhu, X. et al. (2013). Proteomic analysis of heat stress response in leaves of radish (*Raphanus sativus* L.). *Plant Molecular Biology Reporter* 31: 195–203.

Zhang, G.H., Liu, H., Wang, J.J., and Wang, Z.Y. (2014). Effects of thermal stress on lipid peroxidation and antioxidant enzyme activities of the predatory mite, *Neoseiulus cucumeris* (Acari: Phytoseiidae). *Experimental and Applied Acarology* 64: 73–85.

Zhang, S., Fu, W., Li, N. et al. (2015). Antioxidant responses of *Propylaea japonica* (Coleoptera: Coccinellidae) exposed to high temperature stress. *Journal of Insect Physiology* 73: 47–52.

Zou, J., Liu, C., and Chen, X. (2011). Proteomics of rice in response to heat stress and advances in genetic engineering for heat tolerance in rice. *Plant Cell Reports* 30: 2155–2165.

Zou, M., Yuan, L., Zhu, S. et al. (2017). Effects of heat stress on photosynthetic characteristics and chloroplast ultrastructure of a heat-sensitive and heat-tolerant cultivar of wucai (*Brassica campestris* L.). *Acta Physiologiae Plantarum* 39 (1): 30. https://doi.org/10.1007/s11738-016-2319-z.

Źróbek-Sokolnik, A. (2012). Temperature stress and responses of plants. In: *Environmental Adaptations and Stress Tolerance of Plants in the Era of Climate Change* (eds. P. Ahmad and M.N.V. Prasad), 113–134. New York: Springer.

4

Biochemical and Molecular Markers: Unraveling Their Potential Role in Screening Germplasm for Thermotolerance

Ahmed Ismail[1,2], Kareem A. Mosa[1,3], Muna A. Ali[3], and Mohamed Helmy[4]

[1] *Department of Biotechnology, Faculty of Agriculture, Al-Azhar University, Cairo, Egypt*
[2] *Public Health Department, College of Applied Medical Science, Majmaah University, Majmaah City, KSA*
[3] *Department of Applied Biology, College of Sciences, University of Sharjah, Sharjah, UAE*
[4] *The Donnelley Centre, University of Toronto, Toronto, ON, Canada*

4.1 Introduction

Producing sufficient food for rapidly increasing populations is an important challenge particularly as the threat of climate change looms for millions of people (Abberton et al. 2016). As a consequence of global warming, temperature changes are also widespread. Increased temperatures throughout the reproductive stage also significantly influences the growth of plants all over the world (Kumar et al. 2013). Therefore, constant high temperatures affect plant development and lead to morphological, molecular, physiological, and biochemical changes which result in decreasing yield (Figure 4.1). The rapid occurrence of drought and heat stress also has a significant impact on crop productivity (Lesk et al. 2016). Therefore, attaining sizeable yield gains in the major crops is important to meet rising demand. This has been estimated to require a 70% increase in crop production by 2050 (Tilman et al. 2011). In response, several genetic (molecular and biochemical) approaches have been developed to identify thermotolerant plant germplasms (Larkindale et al. 2005).

Heat Stress Tolerance in Plants: Physiological, Molecular and Genetic Perspectives, First Edition.
Edited by Shabir Hussain Wani and Vinay Kumar.

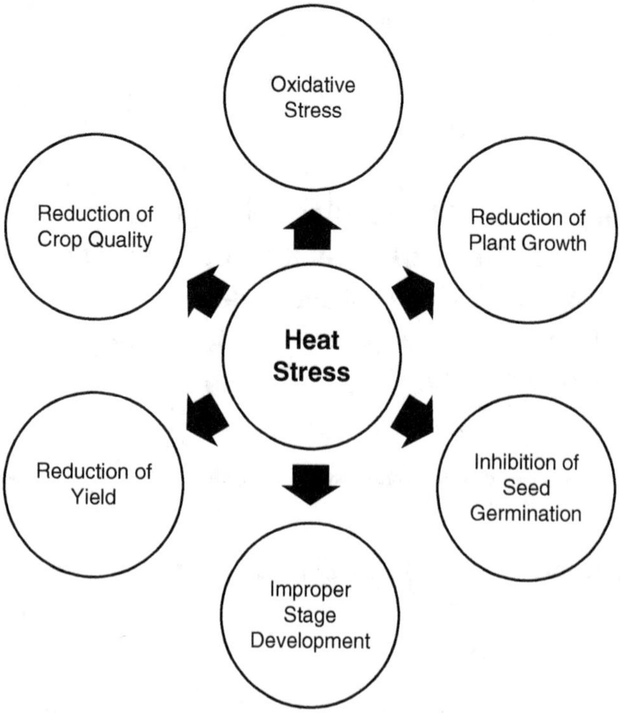

Figure 4.1 Major consequences of the heat stress on plants.

4.2 Types of Markers

The evaluation of genetic variety in crop species is the foundation of any intended improvements. Plant breeders continually attempt to develop improved varieties and/or hybrids in relation to genetic variability. The knowledge of the diversity of germplasms and their relation to selected breeding materials have improved significantly in crop plants. This information is valuable when planning for hybrid and line growth crosses as well as assigning lines to heterotic groups. Plants have both an inherent capability to survive heat above the growth optimal as well as genetically gain tolerance to lethal heat stress (Larkindale et al. 2005). By reviewing and understanding crop genetic diversity and identifying potential candidate genes, the integration of genetic, genomic, and phenotypic information is crucial in confirming the widespread applications of novel genetic discoveries. There are three different types of markers: morphological, molecular, and biochemical (Figure 4.2). We introduce each of them in the following sections.

4.3 Morphological Markers

Morphological markers are related to features like shape, color, size, and surface of the different thermotolerance plant parts. Such features can be used for verities identification. Plants possess several advanced mechanisms to survive under high temperatures (Mosa et al. 2017a,b). These strategies are categorized into long-term morphological evolutionary variations, such as cooling transpiration, alterations in the leaf direction, or alterations in the membrane lipid structure, as well as short-term heat stress, such as avoidance and acclimation mechanisms (Mosa et al. 2017a and b). Plants adapting to heat stress develop appropriate changes to morphological characteristics. Therefore, studying plant morphology in response to high temperature stress often enhances our understanding of plant responses and helps to further characterize molecular tolerance (Wahid et al. 2007). The temperature and length of heat stress resulting in growth and developmental changes differ between plant tissues and growth stages. Therefore, the choice of the right phenotype is important as the gene response to heat stress can differentially express thermotolerance through various tissues and growth stages. Systematic phenotyping methodology, including a variety of heat stress conditions, can preferably increase the probabilities of identifying genes that are involved in potential heat stress responses (Sivasankar et al. 2012).

4.4 Molecular Markers

Many genome-wide approaches are used for evaluating and identifying alleles linked to desirable agronomic qualities in diverse germplasm groups. Single nucleotide polymorphisms (SNPs) and QTL are two approaches related to required

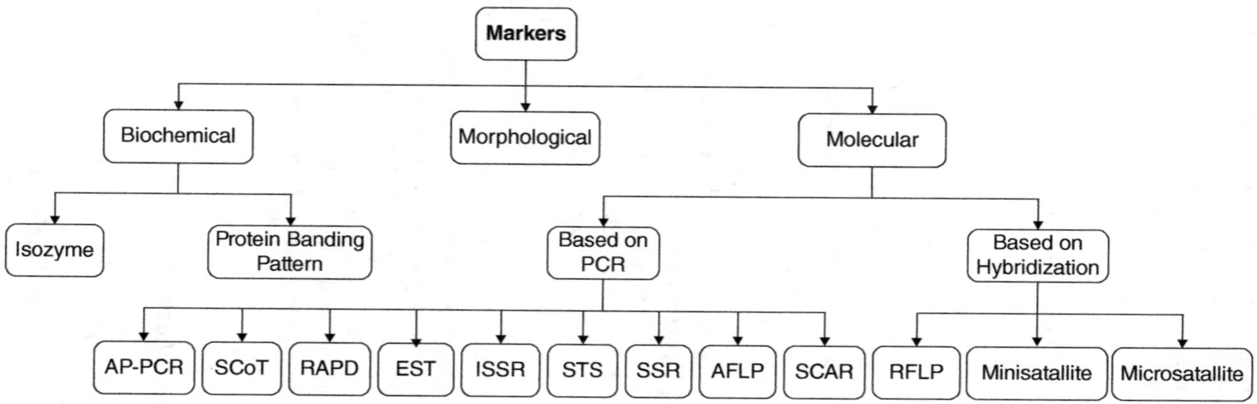

Figure 4.2 Different types of markers.

agronomic traits that have been used to improve crop productivity and resistance. The inclusion of alleles associated with genetic variants as well as their related pathways can provide environmental flexibility along with improving plant yields. Furthermore, SNPs and the discovery of rare alleles provide opportunities to improve plant tolerance to abiotic stresses such as high temperature (heat stress), enhancing their value as resources for speeding up breeding under heat-prone environments (Dwivedi et al. 2017).

Transcriptome analysis using the microarray approach operating cross-hybridization to Affymetrix oligonucleotides and the Rice GeneChip® Genome Array (~33 700 gene) are among the common attempts toward identifying abiotic stress-responsive marker genes. In this case, many have been reported and are highly promising in relation to gene discovery strategies. Many genes involved in response to abiotic stimulus have been reported, including heat stress factors (HSFs) of transcription factors (TFs) involved in plant abiotic stress control (Davey et al. 2009). Microarray data have also revealed the correlation of expression profiles between stress-regulated transcripts and make it easier to predict their cis-regulatory elements. Microarray analysis of two genotypes of *Medicago truncatula* was conducted to compare 384 genes related to salinity and other abiotic stresses. Analysis in the root showed that several genes of the TFs not only responded to the salt stress, but also up-regulated when exposed to heat and cold stress (Gruber et al. 2009). More recently, the TRUNCATULIX warehouse of five integrated databases (gene sequences along with the raw data of microarray expression, normalized datasets, annotations, and complete expression profiling) was built from several *M. truncatula* databases and made accessible at http://lily.cebitec.uni-bielefeld.de/truncatulix/app (Henckel et al. 2009). Thus, this form of analysis allows data mining for a particular species and leads to increased understanding of stress tolerance and response.

PARP1 and *PARP2* are two plant genes that are activated when DNA is damaged following exposure to stress, leading to energy depletion and ATP consumption. The inhibition or down-regulation as a result of expression of these genes contributes to a wide variety of abiotic stresses, such as heat, high light, and drought tolerance (De Block et al. 2005). Improved thermotolerance, photosynthesis, seed production, and plant growth under heat stress were reported in transgenic *Arabidopsis* with wild-type Rubisco activase (RCA), a relative protein (Kurek et al. 2007). The analysis of *mpk4* during stress responses has revealed that the activation of *AtMKK2* leads to activation of *AtMPK4* kinase activity, as well as inducing the mRNA expression of several cold stress marker genes such as *CBF2* and *CBF3* (Kurek et al. 2007).

SNPs represent the evaluation of single-nucleotide differences in genetic sequences variation. Normally, differences are located either by increasing the PCR time of the target DNA regions (within the gene or in a regulatory region in close proximity to the gene) between different individuals followed by fragment sequencing or utilizing methods to analyze differences in fragment conformation and mobility such as single-strand polymorphism conformation (SSCP) or high-resolution melting (HRM). The availability of reference genomes, along with recent approaches such as SNP and short tandem repeat (SSR), allow effective *in silico* exploration by comparing sequence diversity between lines, strains, and genotypes. The enormous volume of data generated by next-generation sequencing (NGS), the genome of a particular crop, and locating polymorphisms related to genes of interest effectively support the development and data mining of molecular markers (Imelfort et al. 2009).

On the other hand, aligning, and short reads clustering produced by NGS platforms is particularly difficult to perform for plants with complex and large genomes, in the absence of a reference genome, and during the abundance of repeated sequences. As such, these challenges complicate the discovery of genome-wide SNPs using NGS in total genomic DNA, requiring the use of specialized approaches in SNP mining (Helmy et al. 2012; El-Metwally et al. 2013). For example, in *Aegilops tauschii* (wild diploid wheat), an improved pipeline protocol for wide detection, was utilized in a genome size of 4.02 GB with 90% repeated sequences, revealing 500 000 SNPs (You et al. 2011). Association maps of SNPs can also be used for genome-wide association (GWA) and mapping as well as genomic selection (GS).

Since sorghum is well adapted to growth in warm conditions, exposure to low temperatures in the early growing stage is a major limitation. Despite most of the existing sorghum germplasms belonging to a tropical source, other landraces from low-temperate regions in China (known as kaoliang) demonstrate a greater seedling potency under cold conditions (Singh 1985). Efforts are also underway in utilizing QTL analysis in early-stage cold performance to develop a population through crossing SRN39 (cold sensitive) and Shan Qui Red (SQR, cold tolerant) in order to transmit required genes from the Chinese landrace into the desired germplasm (Knoll et al. 2008). As a result, two QTL (SBI-03 and SBI-07) were identified for germination under cold temperatures, while a third QTL located on linkage group SBI-01 was associated with seedling development and potency over late and early field planting (Knoll and Ejeta 2008).

4.5 Biochemical Markers

Nowadays, the utilization of biochemical mechanisms and strategies of heat tolerance for improving crop thermotolerance has become essential. Plants have various adaptive, preventative, or acclimatizing mechanisms to deal with high-temperature situations. Through ontogeny, heat stress has an effect on plant growth, though the threshold level fluctuates significantly in different developmental stages. For example, heat can delay or completely inhibit germination at the seed germination stage in accordance with the plant species and stress level involved. In later stages, high temperatures can damage respiration, photosynthesis, water interactions, and membrane stability, as well as moderate hormone levels and the production of primary and secondary metabolites (Mosa et al. 2017a). Improved expression of various types of heat shock and stress-correlated proteins as well as reactive oxygen species (ROS) production are the main plant biochemical reactions to heat stress, leading to oxidative stress (Hasanuzzaman et al. 2012).

Plants perform various mechanisms including transcriptional activation, signaling of chaperones, antioxidant production, membrane stability maintenance, alteration of compatible solutes, as well as the production of calcium-dependent protein kinase (CDPK) and mitogen-activated protein kinase (MAPK).

All these mechanisms are regulated at the molecular level in order for plants to adapt under heat stress. Moreover, major tolerance mechanisms involving antioxidants, osmoprotectants, ion carriers, and additional factors partaking in signaling flows and transcription regulation are activated to balance biochemically induced changes.

Proteomes are linked to different biochemical processes and are produced under the metabolomics platform by several metabolic products (metabolome). Heat shock proteins (HSPs) are biochemical markers that play a pivotal role in the cold tolerance of fruits and vegetables. HSPs are crucial for sustaining membrane characteristics and attributes under low-temperature stress, thanks largely to their chaperone activity. When comparing heat shock with cold shock responses in *Arabidopsis*, different metabolomics demonstrated overlapping of major metabolites in response to both forms of shock. RNA-binding protein *ATGRP7* was identified to be decreased in response to extreme temperatures while levels were increased under low-temperature stress, with significant correlation to glutamine and proline but not raffinose and galactinol (Wienkoop et al. 2008).

4.6 Quantitative Trait Loci for Plant Thermotolerance

Most of the agriculturally important traits in plants, including quality, yield, pest, and disease resistance as well as abiotic stress tolerance (including thermotolerance), are complex traits (Veeresha et al. 2015). Complex traits (also known as quantitative traits or polygenes) are controlled by more than one gene, and the loci that carry one or a cluster of genes controlling a quantitative trait is known as quantitative trait loci (QTLs) (Collard et al. 2005). QTL mapping is considered one of the most important forward genetics strategies to identify and functionally characterize genes under abiotic stress (Mosa et al. 2017c). QTL analysis is accomplished by creating a linkage map of the QTLs using molecular markers to identify loci associated with the trait of interest, known as QTL mapping

(Veeresha et al. 2015). QTL analysis detects the link between the phenotype and genotype in a population having genetic variation (Broman 2009), followed by statistical analysis to detect if there is a significant difference between a group with one genotype and a group with another genotype based on the trait of interest (Sehgal et al. 2016). To detect thermotolerance, several traits has been analyzed such as reproductive heat tolerance (Rahaman et al. 2018; Xu et al. 2017; Ye et al. 2012; Zhu et al. 2017), physiological traits as chlorophyll content and membrane stability (Awlachew et al. 2016), grain yield (Paul et al. 2018), and flower quality (Branham et al. 2017).

There are two main methods that are used for QTL mapping: traditional QTL mapping (linkage analysis) and association mapping (Sehgal et al. 2016). Linkage analysis begins with crossing two parents with different phenotypes in the trait(s) of interest to allow for the analysis of more polymorphisms in the generated population (Dhingani et al. 2015). There are several types of mapping populations that can be developed from heterozygous F1 self-pollinating plants (Table 4.1). Different types of mapping populations were used to determine QTLs related to thermotolerance. The choice of the method used depends on the plant species, type of marker, and the trait that will be detected (Sehgal et al. 2016). Since cross-pollinating plants cannot tolerate inbreeding, the mapping populations are produced by crossing a heterozygous parent with a haploid or homozygous parent (Collard et al. 2005).

After generating the mapping population, the phenotype of the population is screened for the trait of interest (Sehgal et al. 2016). Subsequently, the molecular data of the population (along with the parents) will be generated based on the molecular marker that will be analyzed. There are different types of molecular markers that are commonly used in QTLs (Figure 4.2). Generally, the markers are divided into two types: dominate markers and codominant markers. Codominant markers, for example RFLP (restriction fragment length polymorphism) and SSR, can differentiate between homozygotes and heterozygotes, while dominant markers, for example RAPD (random amplified polymorphic DNA) and AFLP (amplified fragment length polymorphism), cannot. The choice of the utilized DNA marker depends on the plant species, number of available markers, as

Table 4.1 The different types of mapping populations used to study QTLs.

Population	Procedure	Advantages	Limitations	Plant example (crossed cultivars)	References
F2 population	Selfing F1 hybrid	Easy to construct, require short time to produce	For self-pollinating plants, temporary populations due to the high heterozygosity	*Solanum lycopersicum* (Nagcarlang (LA2661) × NCHS-1 (LA3847))	Collard et al. (2005) and Xu et al. (2017)
Backcross (BC) population	F1 plants are backcrossed to one of the parents			*Oryza sativa* (japonica variety Sasanishiki × indica elite variety Habataki)	Collard et al. (2005) and Zhu et al. (2017)
Double haploid lines (DHLs)	Pollen of F1 plants is treated to cause chromosome doubling and restore the diploid condition	Produce homozygous true breeding lines without any genetic change, require short time	Only possible in plants that are amenable to tissue culture (e.g. cereal species)	*Brassica oleracea* (USVL138 × USVL108), *Triticum aestivum L.* (Hanxuan 10 × Lumai)	Sehgal et al. (2016) and Branham et al. (2017)
Recombinant inbred lines (RILs)	Inbreeding from individual F2 plants after 6–8 generations	Produce homozygous true breeding lines without any genetic change	For self-pollinating plants, long time to produce (6–8 generations)	*Cicer arietinum* (ICC 4567 × ICC 15614)	Paul et al. (2018) and Dhingani et al. (2015)
Near isogenic lines (NILs)	Selfing of selected individuals from BC7F1 in order to make lines that differ only in a single QTL			*O. glaberrima* (CG14 × WYJ)	Dhingani et al. (2015) and Li et al. (2015)

The advantages and limitations of each method is presented in this table with an example of plant species in which the heat stress tolerance QTLs were detected.

well as the advantages and limitations which are associated with each marker (Collard et al. 2005).

Linkage maps will be generated based on the positions of the markers along the chromosome, which depends on the frequency of the recombinant genotypes (Sehgal et al. 2016). The odds ratio (linkage versus no linkage) is calculated using the LOD score. In addition, there are several computer packages that can be used to construct the linkage map, such as JoinMap, MAPMAKER/EXP, GMENDEL, LINKAGE, and Map Manager QTX (Dhingani et al. 2015). To detect QTLs and estimate their locations along the genome, four methods are commonly used: single-marker analysis, interval mapping by maximum likelihood, interval mapping by regression, and composite interval mapping (Table 4.2). There are more than 100 software packages that can be used for QTL mapping and linkage analysis, such as MapMaker/QTL, MQTL, PlabQTL, and QTL cartographer (Sehgal et al. 2016).

QTL mapping has been widely used in breeding programs. In addition, there are many examples of QTL evaluation that are associated with heat stress tolerance in plants (Table 4.3). Nevertheless, there are several limitations associated with linkage methods. The precision of linkage methods is limited because only two parents are utilized, which reduces the number of polymorphisms in the population. In addition, it is tedious and time-consuming in terms of generation of the mapping populations and results analysis. Moreover, this method is not applicable for some plant species that have long generation times (e.g. forest trees).

To reduce these limitations, association mapping was developed. Instead of crossing two parents, association mapping analyzes natural populations (or germplasm collections). This is more time efficient, surveys larger and more representative gene pools, and provides finer mapping resolution (Sehgal et al. 2016). This method was used in detection of heat tolerance QTLs in *Brassica napus* (canola), where plants from a germplasm resource were used. Genotype-based sequencing was done to sequence 37 539 SNP markers, and the QTL regions were shown to be associated with male sterility, embryo abortion reducing pollen development, pollen abortion, pod development, and flowering.

Table 4.2 The commonly used methods to detect QTLs and estimate their locations.

QTL detection method	Definition
Single marker analysis	Studying a single genetic marker at a time The t-test or F-test is used to the significance of the linkage between the marker and the QTL Does not require a complete genetic map construction
Interval mapping by maximum likelihood (standard interval mapping)	The most common method for QTL analysis Tests the possibility of QTL localization between two markers Uses maximum likelihood by searching for QTL parameters that give the best approximation for quantitative trait distribution
Interval mapping by regression	A simplified maximum likelihood method Probabilities estimated from the flanking markers are used to indicate QTL genotypes Then, the phenotypes are regressed on QTL genotypes
Composite interval mapping	A method that removes the bias caused by linked QTLs Accounts for the variance from other QTLs by including the partial regression coefficients from markers in other regions

These analyses are done via different software packages specialized for QTL analysis.
Source: Adapted from Sehgal et al. (2016).

Based on the QTL results, the molecular mechanisms underlying plant heat tolerance can be elucidated. Furthermore, it can be applied for germplasm enhancement and marker assisted breeding in order to obtain thermotolerant species of crop plants. Moreover, after verifying the QTLs, they can be genetically modified in different crop plants to improve their thermotolerance abilities. For example, the overexpression of thermotolerance-1 (TT1) QTL identified in *Oryza glaberrima* (African rice) led to improved thermotolerance in rice, *Arabidopsis,* and *Festuca elata.* This may help in increasing crop security under global warming. In addition, QTL mapping combined with genome-wide miRNA profiling can enhance the identification of miRNA and their responsible traits (Liu et al.

Table 4.3 QTL mapping studies in different plants under heat stress to detect QTLs associated with thermotolerance.

Plant species	Evaluated traits	Genotype marker	Linkage map construction program	QTL analysis program	QTL detection method	Trait (number of QTLs)	References
Solanum lycopersicum	Inflorescence number (IN), flowers per inflorescence (FPI), anther length (AL), style length (SL), style protrusion (SP), pollen viability (PV), pollen number (PN), female fertility (FF)	SNP	Rqtl	Rqtl	SIM, CIM	PV (1), PN (1), FPI (1), IN (2), SP (2), AL (3), SL (3)	Xu et al. (2017)
Aegilops speltoides	Cell membrane thermostability (CMS), chlorophyll content at anthesis (CCA), acquired thermotolerance (ATT), canopy temperature (CT), stay green (SG), vegetative period (VP).	SSR	IciMapping	IciMapping	SMA, RSTEP-LRT-ADD	CCA (1), CT (1), ATT (1), SG (4)	Awlachew et al. (2016)
Cicer arietinum L.	Filled pods per plot (FPod), total number of seeds per plot (TS), grain yield per plot (GY) and % pod setting (%PodSet), visual score on podding behavior (VS)	SNP	JoinMap 4.	QTL Cartographer-V 2.5	CIM	Fpod (2), %PodSet (2), GY (2), TS (1), VS (1)	Paul et al. (2018)

(Continued)

Table 4.3 (Continued)

Plant species	Evaluated traits	Genotype marker	Linkage map construction program	QTL analysis program	QTL detection method	Trait (number of QTLs)	References
Winter wheat	75% green (75% G), 50% green (50% G), 25% green (25% G), Percentage chlorophyll content (SPAD), maximum rate of senescence (MRS), time to maximum rate of senescence (TMRS), percent greenness at maximum senescence (PGMS), Chlorophyll fluorescence (Fv/Fm)	AFLP, SSR	Mapmaker/ EXP for UNIX version 3	QTL Cartographer version 2.0	CIM	75% G (3), 50% G (2), 25% G (2), MRS (2), TMRS (3), PGMS (2), Fv/fm (1)	Kiran et al. (2016)
Brassica oleracea var. italica	Broccoli head quality based on color, size, shape, smoothness, firmness, and bead size and uniformity	SNP	Rqtl	Rqtl	IMR	(5)	Branham et al. (2017)

SNP, single nucleotide polymorphism; SSR, short tandem repeat; AFLP, amplified fragment length polymorphism; SIM, standard interval map; CIM, composite interval map; SMA, single marker analysis; IMR, interval mapping by regression.

2017). This can help to facilitate the molecular breeding of heat tolerant cultivars.

4.7 Plant Metabolites Under Heat Stress

There are many internal and external factors that contribute to the final phenotype of the living organism. Internal factors include DNA, RNA, proteins, and metabolites, while external factors are related to environmental influences. The integration of genomic, transcriptomic, proteomic, and metabolomic studies are vital for the examination of how plants interact and respond to environmental stresses (Helmy et al. 2011). The metabolomics approach is used to obtain extensive information that can be further utilized in the discovery of genes and pathways. As plant metabolites are abundant and diverse, the detection method as well as the employed instruments largely affect the result (Kumar et al. 2018). Thus, there is no single approach to detecting all metabolites (Hong et al. 2016). Metabolic profiling methods include capillary electrophoresis-liquid chromatography-mass spectrometry (CE-MS), gas chromatography (GC), mass spectrometry (MS), nuclear magnetic resonance (NMR), flow injection analysis coupled to MS (FIA/MS) and Fourier transform infrared spectroscopy (FTIR) (Arbona et al. 2013; Mosa et al. 2017d).

During stress conditions, the metabolic profile of the plant is altered in order to respond appropriately. Metabolomics have been utilized to elucidate alterations in the metabolic profile of plants during such situations. The metabolite composition of the plant is based on the gene expression profile. Gene expression regulates the activated metabolic pathways, which in turn alters the metabolic profile. In addition, the produced metabolites are also regulated by other molecules from unrelated pathways (Figure 4.3) (Arbona et al. 2013). Gene expression is a complex process and may involve the integration of numerous genes, epigenetic, and environmental factors. Several QTLs have been linked to metabolite abundance in plants. They are referred to as metabolomic QTL (mQTL) (Matros et al. 2017), and may be used to identify genotypes that tolerate environmental stresses.

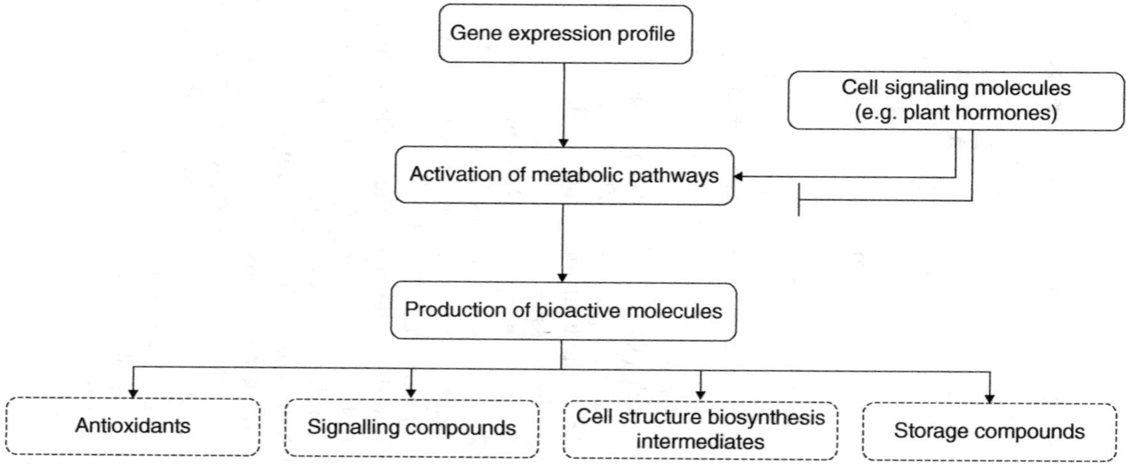

Figure 4.3 The metabolic profile of the plant is based on the gene expression pattern and the cell signaling molecules. The gene expression profile activates metabolic pathways, and cell signaling molecules activate or inhibit the steps of the metabolic pathway.

Generally, there are two main types of plant metabolites: primary metabolites, which are essential for plant growth and development and are conserved between species, and secondary metabolites of various types, which are not necessary for growth but are important for protection during stress conditions. In response to heat stress, several types of primary and secondary metabolites accumulate in the plant (Figure 4.4). Among these are osmoprotectants (including proline, glucose, fructose, galactinol, raffinose, and glycine betaine) that play an important role in heat stress tolerance. The accumulated compounds depend on the type of plant species, tested plant part, growth stage, heat stress level and duration, and the method used for metabolic profiling (Table 4.4) (Bita and Gerats 2013).

Combining transcriptomic and metabolomic data will greatly assist in improving the accuracy of metabolic profiling. This will be facilitated by developing databases that combine the transcriptome and the metabolome (Hasanuzzaman et al. 2013), and eventually lead to a better understanding of the mechanisms underlying stress response in plants and help in improving crop plant breeding.

4.8 Antioxidant Enzymes and Heat Stress

During the heat stress response in plants, oxidative stress occurs as a secondary stress (Qu et al. 2013). Oxidative stress leads to the accumulation of different types of ROS, most commonly singlet oxygen ($^1O_2$), superoxide radical ($O^{2\bullet}$), hydrogen peroxide (H_2O_2), and hydroxyl radical ($OH^\bullet$) (Hasanuzzaman et al. 2013). Upon ROS production, cellular and molecular damage occurs in the plant (Figure 4.5) (Ahmad et al. 2008). ROS signals are transmitted by HSFs to induce the MAPK signaling pathway, which will lead to the activation of ZAT (zinc finger of *Arabidopsis thaliana*) and WRKY TFs. ZAT and WRKY will induce the expression of several antioxidant enzymes that will scavenge ROS (Qu et al. 2013).

Antioxidant enzymes have important roles during heat stress tolerance in plants. They maintain cell homeostasis and scavenge ROS that are produced during plant stress conditions. Thus, they are considered a defense system against oxidative

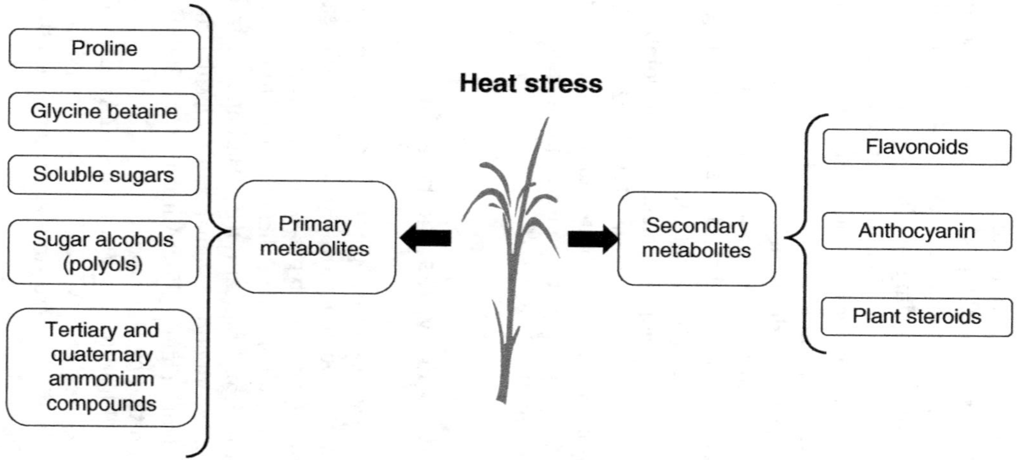

Figure 4.4 Metabolite production in plants under heat stress. During heat stress, several types of primary and secondary metabolites accumulate in the plants.

Table 4.4 The differentially produced metabolite under heat stress in different plant species.

Plant species	Tested plant part	Heat stress	Metabolomic profiling method	Affected metabolite groups (level compared to control[a)])	References
Glycine max L.	Leaf	43/35 °C light/dark for 7 d	LC/MS & GC/MS	Secondary metabolites (↓), aromatic amino acids (↓), fatty acids (↓), TCA cycle metabolites (↓), glycolytic metabolites (↓), pentose phosphate pathway metabolites (↓), peptides (↓), purine and pyrimidine metabolites (↓)	Das et al. (2017)
Triticum turgidum L. genotype: Primadur	Immature seeds	37/17 °C light/dark for 5 d	GC/MS	Amino acids (↑) exc. Ser, Val, Thr, β-Ala, Trp (–) and Ile (↓), N-compounds (↑) exc. Spermidine (–) organic acid (↑) exc. Nicotinic acid (↓), quinic acid(↓) shikimic acid (–), sugars (↑) exc. Glucose and raffinose (–), sugar alcohols (–) exc. Myo-inositol (↑)	de Leonardis et al. (2015)
	Mature seeds			Amino acids (↑), N-compounds (↑), organic acid (↑) exc. 3 PGA (↓), sugars (↑), sugar alcohols (↑)	
Triticum turgidum L. genotype: T1303	Immature seeds			Amino acids (–) exc. Ala and Trp (↑), N-compounds (–), organic acids (–) exc. Citric acid, fumaric acid, malic acid, oxalic acid (↓), sugars (–) exc. Raffinose (↓), sugar alcohol (–) exc. Mannitol (↓)	

(Continued)

Table 4.4 (Continued)

Plant species	Tested plant part	Heat stress	Metabolomic profiling method	Affected metabolite groups (level compared to control[a])	References
	Mature seeds			Amino acids (↓) exc. Val, Ala, Ile, and Leu (−), N-compounds (−) exc. Spermidine (↓), organic acids (−) exc. Shikimic acid (−) and 3 PGA (↑), sugars (↓), sugar alcohol (↓) exc. Glycerol (−)	
Nicotiana langsdorffii	Whole plant	50 °C for 2 h	HPLC and HRMS	Sulfolipids (↑), galactolipids (↑), phosphatidylethanolamine (↑), acylglycerols (↑), acylsugars (↑), steroidal glycolalkaloids (↑)	Scalabrin et al. (2015)
Zea mays L.	Leaf	Planted at the start of the dry season	GC/MS	Phe (↑), Ala (↑), GABA (↑), threonate (↑), xylose (↑), galactinol (↑), Ile (↓), glycerol (↓), malate (↓), glycerate (↓)	Obata et al. (2015)

a) The relative production compared to control plants: increased (↑), decreased (↓), remained the same (−).

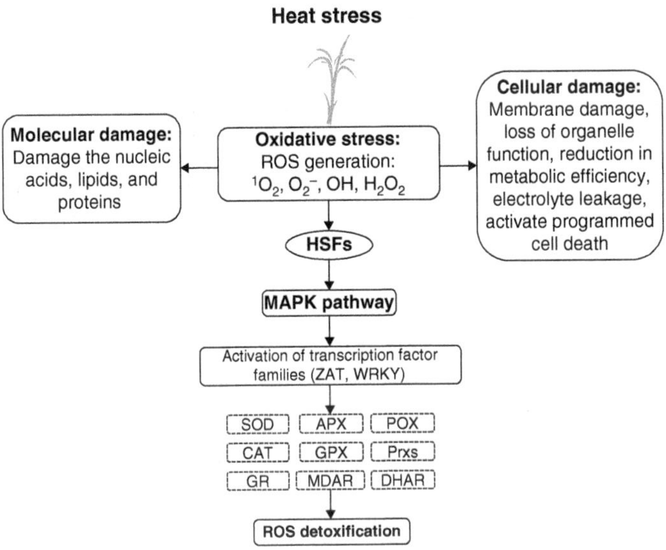

Figure 4.5 The effect of heat stress on the plant. Heat-stress-induced oxidative stress as a secondary stress which causes the production ROS molecules that will lead to molecular and cellular damage. ROS is sensed by HSFs which activates the MAPK signaling pathway, which will lead eventually to the synthesis of antioxidant enzymes. SOD (superoxide dismutase), APX (ascorbate peroxidase), CAT (catalase), GPX (glutathione peroxidase), POX (guaiacol peroxidase), Prxs (peroxiredoxins), MDAR (monodehydroascorbate reductase), DHAR (dehydroascorbate reductase), and GR (glutathione reductase) enzymes.

stress (Caverzan et al. 2016). The expression of antioxidant enzymes is affected by temperature. Generally, the activities of these enzymes increase with temperature, with different antioxidant enzymes having different temperature ranges (Table 4.5). In addition, the activity of antioxidant enzymes depends on the degree of heat tolerance of a particular plant variety, the growth stage, and the growing season (Hasanuzzaman et al. 2013). There are over 150 genes in plants that encode ROS-detoxifying or ROS-producing enzymes (Ahmad et al. 2008). In the last decades, considerable advances have been made in the discovery of genes responsible for plant stress tolerance. These significant findings have allowed the generation of stress-tolerant transgenic plants (Table 4.6).

Table 4.5 The relative expression level of antioxidant enzymes in different plant species under heat stress compared to control plants.

Plant species	Treatment	Antioxidant enzyme and its expression level[a)]	References
Brassica juncea	45 °C for 4.5 h	POD (↑), CAT (↓)	Wilson et al. (2014)
Solanum lycopersicum L.	42 °C for 6 h	SOD (↑), CAT (↑), APX (↑), GR (↑)	Zhou et al. (2014)
Sorghum bicolor L.	40 °C for 6 h	SOD (↑), POD (↑), CAT (↑)	Gosavi et al. (2014)
Oryza sativa L.	35/30 °C light/dark for 10 d	SOD (↑), CAT (↑), APX (↑), GR (↑)	Nayyar et al. (2014)
	42/37 °C light/dark for 10 d	SOD (↓), CAT (↓), APX (↓), GR (↓)	
Agrostis stolonifera	35/30 °C light/dark for 21 d	SOD (↓), CAT (↓), POD (↓), APX (–), DHAR (↓), GR (–)	Ding et al. (2016)
	35/30 °C light/dark for 28 d	SOD (↓), CAT (↓), POD (↓), APX (↓), DHAR (↓), GR (↑)	
	35/30 °C light/dark for 35 d	SOD (↓), CAT (↓), POD (↓), APX (↓), DHAR (↓), GR (–)	

a) The relative expression compared to control plants: (↑) increased expression, (↓) decreased expression, or (–) no significant difference.

4.9 Conclusion

Detection of thermotolerance in germplasms is a vital step in developing heat-tolerant plants and improving plant-breeding processes. Several types of markers are utilized to evaluate plant thermotolerance: morphological, biochemical, and molecular markers. Morphological and biochemical markers depend on the plant growth stage and environmental conditions, and they may be limited in number. Besides, biochemical markers are affected by the extraction methodologies and plant tissue. Plant breeders widely use morphological and biochemical markers, despite their limitations. Molecular markers are the

Table 4.6 Examples of transgenic plants with overexpressed genes encoding antioxidants and their effects on increasing the thermotolerance of the target species.

Gene	Encoded antioxidant enzyme	Promoter	Native species	Target species	Effect on target species	References
BcAPX	Ascorbate peroxidase (APX1)	Ubiquitin	*Brassica campestris* L.	*Arabidopsis thaliana* L.	Higher APX activity under high temperature, higher chlorophyll content and germination rate, lower malondialdehyde and H_2O_2	Chiang et al. (2015)
BoCAT1 and *BoCAT2*	Catalase (CAT)	Ubiquitin	*Brassica oleracea*	*Arabidopsis thaliana* L.	Higher CAT activity, higher CAT accumulation, lower level of heat induced H_2O_2	Chiang et al. (2014)
cAPX	Cytosolic ascorbate peroxidase (APX)	lac	*Pisum sativum* L.	*Lycopersicon esculentum*	Reduced electrolyte leakage from leaf disks, higher APX activity under heat stress	Wang et al. (2006)
pSSA-K (chimeric gene construct)	Cu/Zn SOD and APX	SWPA2 (oxidative stress inducible)	Cu/Zn SOD gene source: *Manihot esculenta* Crantz. APX gene source: *Pisum sativum*	Two transgenic potato lines (SSA9 and SSA11)	Photosynthetic activity decreased by only 6% compared to 29% decrease in nontransgenic plants	Tang et al. (2006)
HvAPX1	Peroxisomal type ascorbate peroxidase	35S RNA promoter of the cauliflower mosaic virus	*Hordeum vulgare* cv. Haruna-nijyo	*Arabidopsis thaliana*	Higher percentages of green part over the total	Shi et al. (2001)

most commonly used markers, owing to their abundance, independence on plant growth stage, and environmental conditions, precision, time-saving, and reliability. Molecular markers are utilized in QTL mapping to locate the genes that affect the traits of interest (Kordrostami and Rahimi 2015). Molecular markers, metabolic markers, and biochemical markers (antioxidants) are utilized to understand thermotolerance associated genes and metabolites. This served as the basis of improving the breading of thermotolerant plant varieties.

References

Abberton, M., Batley, J., Bentley, A. et al. (2016). Global agricultural intensification during climate change: a role for genomics. *Plant Biotechnology Journal* 14 (4): 1095–1098.

Ahmad, P., Sarwat, M., and Sharma, S. (2008). Reactive oxygen species, antioxidants and signaling in plants. *Journal of Plant Biology* 51 (3): 167–173.

Arbona, V., Manzi, M., de Ollas, C., and Gómez-Cadenas, A. (2013). Metabolomics as a tool to investigate abiotic stress tolerance in plants. *International Journal of Molecular Sciences* 14 (3): 4885–4911.

Awlachew, Z.T., Singh, R., Kaur, S. et al. (2016). Transfer and mapping of the heat tolerance component traits of *Aegilops speltoides* in tetraploid wheat *Triticum durum*. *Molecular Breeding* 36 (6): 78.

Bita, C.E. and Gerats, T. (2013). Plant tolerance to high temperature in a changing environment: scientific fundamentals and production of heat stress-tolerant crops. *Frontiers in Plant Science* 4: 273.

Branham, S.E., Stansell, Z.J., Couillard, D.M., and Farnham, M.W. (2017). Quantitative trait loci mapping of heat tolerance in broccoli (*Brassica oleracea* var. italica) using genotyping-by-sequencing. *Theoretical and Applied Genetics* 130 (3): 529–538.

Broman, K.W. (ed.) (2009). Single-QTL analysis. In: *A Guide to QTL Mapping with R/qtl*, 75–133. New York: Springer.

Caverzan, A., Casassola, A., and Patussi Brammer, S. (2016). Reactive oxygen species and antioxidant enzymes involved in plant tolerance to stress. In: *Abiotic and Biotic Stress in Plants: Recent Advances and Future Perspectives* (eds. A.K. Shanker and C. Shanker). InTech Open.

Chiang, C.M., Chen, S.P., Chen, L.F.O. et al. (2014). Expression of the broccoli catalase gene (BoCAT) enhances heat tolerance in transgenic *Arabidopsis*. *Journal of Plant Biochemistry and Biotechnology* 23 (3): 266–277.

Chiang, C.M., Chien, H.L., Chen, L.F.O. et al. (2015). Overexpression of the genes coding ascorbate peroxidase from *Brassica campestris* enhances heat tolerance in transgenic *Arabidopsis thaliana*. *Biologia Plantarum* 59 (2): 305–315.

Collard, B.C.Y., Jahufer, M.Z.Z., Brouwer, J.B., and Pang, E.C.K. (2005). An introduction to markers, quantitative trait loci (QTL) mapping and marker-assisted selection for crop improvement: the basic concepts. *Euphytica* 142 (1-2): 169–196.

Das, A., Rushton, P., and Rohila, J. (2017). Metabolomic profiling of soybeans (*Glycine max* L.) reveals the importance of sugar and nitrogen metabolism under drought and heat stress. *Plants* 6 (2): 21.

Davey, M.W., Keulemans, J., Vanholme, B. et al. (2009). Assessing the use of heterologous oligonucleotide microarrays for transcriptomics in a non-model species; application to the study of drought stress in Musa. *Acta Horticulturae* 839: 607–614.

De Block, M., Verduyn, C., De Brouwer, D., and Cornelissen, M. (2005). Poly(ADP-ribose) polymerase in plants affects energy homeostasis, cell death and stress tolerance. *The Plant Journal* 41 (1): 95–106.

Dhingani, R.M., Umrania, V.V., Tomar, R.S. et al. (2015). Introduction to QTL mapping in plants. *Annals of Plant Sciences* 4: 1072–1079.

Ding, J., Lu, G., and Li, Y. (2016). Interactive effects of selected pharmaceutical mixtures on bioaccumulation and biochemical status in crucian carp (*Carassius auratus*). *Chemosphere* 148: 21–31.

Dwivedi, S.L., Scheben, A., Edwards, D. et al. (2017). Assessing and exploiting functional diversity in germplasm pools to enhance abiotic stress adaptation and yield in cereals and food legumes. *Frontiers in Plant Science* 8 (August): 1–30.

El-Metwally, S., Hamza, T., Zakaria, M., and Helmy, M. (2013). Next-generation sequence assembly: four stages of data processing and computational challenges. *PLoS Computational Biology* 9 (12): e1003345.

Gosavi, G.U., Jadhav, A.S., Kale, A.A. et al. (2014). Effect of heat stress on proline, chlorophyll content, heat shock proteins and antioxidant enzyme activity in sorghum (*Sorghum bicolor*) at seedlings stage. *Indian Journal of Biotechnology* 13: 356–363.

Gruber, V., Blanchet, S., Diet, A. et al. (2009). Identification of transcription factors involved in root apex responses to salt stress in *Medicago truncatula*. *Molecular Genetics and Genomics* 281 (1): 55–66.

Hasanuzzaman, M., Hossain, M.A., da Silva, J.A.T., and Fujita, M. (2012). Plant response and tolerance to abiotic oxidative stress: antioxidant defense is a key factor. In: *Crop Stress and Its Management: Perspectives and Strategies* (eds. B. Venkateswarlu, A.K. Shanker, C. Shanker and M. Maheswari), 261–315. Dordrecht, The Netherlands: Springer.

Hasanuzzaman, M., Nahar, K., Alam, M.M. et al. (2013). Physiological, biochemical, and molecular mechanisms of heat stress tolerance in plants. *International Journal of Molecular Sciences* 14 (5): 9643–9684.

Helmy, M., Tomita, M., and Ishihama, Y. (2011). OryzaPG-DB: rice proteome database based on shotgun proteogenomics. *BMC Plant Biology* 11 (1): 63.

Helmy, M., Sugiyama, N., Tomita, M., and Ishihama, Y. (2012). The rice proteogenomics database OryzaPG-DB: development, expansion, and new features. *Frontiers in Plant Science* 3 (April): 65.

Henckel, K., Runte, K.J., Bekel, T. et al. (2009). TRUNCATULIX: a data warehouse for the legume community. *BMC Plant Biology* 9: 1–11.

Hong, Y., Zhang, H., Huang, L. et al. (2016). Overexpression of a stress-responsive NAC transcription factor gene *ONAC022* improves drought and salt tolerance in rice. *Frontiers in Plant Science* 7 (January): 1–19.

Imelfort, M., Duran, C., Batley, J., and Edwards, D. (2009). Discovering genetic polymorphisms in next-generation sequencing data: review article. *Plant Biotechnology Journal* 7 (4): 312–317.

Kiran, G.S., Ninawe, A.S., Lipton, A.N. et al. (2016). Rhamnolipid biosurfactants: evolutionary implications, applications and future prospects from untapped marine resource. *Critical Reviews in Biotechnology* 36 (3): 399–415. https://doi.org/10.3109/07388551.2014.979758.

Knoll, J. and Ejeta, Æ.G. (2008). Marker-assisted selection for early-season cold tolerance in sorghum: QTL validation across populations and environments. *Theoretical and Applied Genetics* 116 (4): 541–553.

Knoll, J., Gunaratna, N., and Ejeta, G. (2008). QTL analysis of early-season cold tolerance in sorghum. *Theoretical and Applied Genetics* 116 (4): 577–587.

Kordrostami, M., and Rahimi, M. (2015) Molecular markers in plants: concepts and applications. Genetics in the Third Millennium. 13, 4024–4031.

Kumar, S., Kumari, P., Kumar, U. et al. (2013). Molecular approaches for designing heat tolerant wheat. *Journal of Plant Biochemistry and Biotechnology* 22 (4): 359–371.

Kumar, A., Mosa, K.A., Ji, L. et al. (2018). Metabolomics-assisted biotechnological interventions for developing plant-based functional foods and nutraceuticals. *Critical Reviews in Food Science and Nutrition* 58 (11): 1791–1807.

Kurek, I., Chang, T.K., Bertain, S.M. et al. (2007). Enhanced thermostability of *Arabidopsis* Rubisco activase improves photosynthesis and growth rates under moderate heat stress. *Plant Cell* 19 (10): 3230–3241.

Larkindale, J., Hall, J.D., Knight, M.R. et al. (2005). Heat stress phenotypes of *Arabidopsis* mutants implicate multiple signaling pathways in the acquisition of thermotolerance. *Plant Physiology* 138 (2): 882–897.

de Leonardis, A.M., Fragasso, M., Beleggia, R. et al. (2015). Effects of heat stress on metabolite accumulation and composition, and nutritional properties of durum wheat grain. *International Journal of Molecular Sciences* 16, 16 (12, 12): 30382, 30382–30404, 30404.

Lesk, C., Rowhani, P., and Ramankutty, N. (2016). Influence of extreme weather disasters on global crop production. *Nature* 529 (7584): 84–87.

Li, X., Yang, Y., Henry, R.J. et al. (2015). Plant DNA barcoding: from gene to genome. *Biological Reviews of the Cambridge Philosophical Society* 90 (1): 157–166.

Liu, Q., Yang, T., Yu, T. et al. (2017). Integrating small RNA sequencing with QTL mapping for identification of miRNAs and their target genes associated with heat tolerance at the flowering stage in rice. *Frontiers in Plant Science* 8: 43.

Matros, A., Liu, G., Hartmann, A. et al. (2017). Genome-metabolite associations revealed low heritability, high genetic complexity, and causal relations for leaf metabolites in winter wheat (*Triticum aestivum*). *Journal of Experimental Botany* 68 (3): 415–428.

Mosa, K.A., Ismail, A., and Helmy, M. (2017a). Introduction to Plant Stresses. In: *Plant Stress Tolerance*, 1–19. Cham, Switzerland: Springer.

Mosa, K.A., Ismail, A., and Helmy, M. (2017b). Omics and system biology approaches in plant stress research. In: *Plant Stress Tolerance*, 21–34. Cham, Switzerland: Springer.

Mosa, K.A., Ismail, A., and Helmy, M. (2017c). Functional genomics combined with other Omics approaches for better understanding abiotic stress tolerance in plants. In: *Plant Stress Tolerance*, 55–73.

Mosa, K.A., Ismail, A., and Helmy, M. (2017d). Omics approaches to understand biotic stresses: a case study on plant parasitic nematodes. In: *Plant Stress Tolerance*, 35–54.

Nayyar, H., Kaur, R., Kaur, S., and Singh, R. (2014). γ-Aminobutyric acid (GABA) imparts partial protection from heat stress injury to rice seedlings by improving leaf turgor and upregulating osmoprotectants and antioxidants. *Journal of Plant Growth Regulation* 33 (2): 408–419.

Obata, T., Witt, S., Lisec, J. et al. (2015). Metabolite profiles of maize leaves in drought, heat and combined stress field trials reveal the relationship between metabolism and grain yield. *Plant Physiology* 169 (4): 2665–2683.

Paul, P.J., Samineni, S., Thudi, M. et al. (2018). Molecular mapping of QTLs for heat tolerance in chickpea. *International Journal of Molecular Sciences* 19 (8): 2166.

Qu, A.L., Ding, Y.F., Jiang, Q., and Zhu, C. (2013). Molecular mechanisms of the plant heat stress response. *Biochemical and Biophysical Research Communications* 432 (2): 203–207.

Rahaman, M., Mamidi, S., and Rahman, M. (2018). Genome-wide association study of heat stress-tolerance traits in spring-type *Brassica napus* L. under controlled conditions. *The Crop Journal* 6 (2): 115–125.

Scalabrin, E., Radaelli, M., Rizzato, G. et al. (2015). Metabolomic analysis of wild and transgenic *Nicotiana langsdorffii* plants exposed to abiotic stresses: unraveling metabolic responses. *Analytical and Bioanalytical Chemistry* 407 (21): 6357–6368.

Sehgal, D., Singh, R., and Rajpal, V.R. (2016). Quantitative trait loci mapping in plants: concepts and approaches. In: *Molecular Breeding for Sustainable Crop Improvement* (eds. V.R. Rajpal, S.R. Rao and S.N. Raina), 31–59. Springer.

Shi, W.M., Muramoto, Y., Ueda, A., and Takabe, T. (2001). Cloning of peroxisomal ascorbate peroxidase gene from barley and enhanced thermotolerance by overexpressing in *Arabidopsis thaliana*. *Gene* 273 (1): 23–27.

Singh, S.P. (1985). Sources of cold tolerance in grain sorghum. *Canadian Journal of Plant Science* 65 (2): 251–257.

Sivasankar, S., Williams, R.W., and Greene, T.W. (2012). Abiotic stress tolerance in plants: an industry perspective. In: *Introduction to Plant Abiotic Strees Response* (eds. N. Tuteja, S.S. Gill, A.F. Tiburcio and R. Tuteja). Wiley.

Tang, L., Kwon, S.Y., Kim, S.H. et al. (2006). Enhanced tolerance of transgenic potato plants expressing both superoxide dismutase and ascorbate peroxidase in chloroplasts against oxidative stress and high temperature. *Plant Cell Reports* 25 (12): 1380–1386.

Tilman, D., Balzer, C., Hill, J., and Befort, B.L. (2011). Global food demand and the sustainable intensification of agriculture. *Proceedings of the National Academy of Sciences* 108 (50): 20260–20264.

Veeresha, B.A., Naik, R., Chetti, M.B. et al. (2015). QTL mapping in crop plants: principles and applications. *International Journal of Development Research* 5: 2961–2965.

Wahid, A., Gelani, S., Ashraf, M., and Foolad, M.R. (2007). Heat tolerance in plants: an overview. *Environmental and Experimental Botany* 61 (3): 199–223.

Wang, Y., Wisniewski, M., Meilan, R. et al. (2006). Transgenic tomato (*Lycopersicon esculentum*) overexpressing cAPX exhibits enhanced tolerance to UV-B and heat stress. *Journal of Applied Horticulture* 8: 87–90.

Wienkoop, S., Morgenthal, K., Wolschin, F. et al. (2008). Integration of metabolomic and proteomic. *Molecular and Cellular Proteomics* 7 (9): 1725–1736.

Wilson, R.A., Sangha, M.K., Banga, S.S. et al. (2014). Heat stress tolerance in relation to oxidative stress and antioxidants in *Brassica juncea*. *Journal of Environmental Biology* 35 (2): 383.

Xu, J., Driedonks, N., Rutten, M.J.M. et al. (2017). Mapping quantitative trait loci for heat tolerance of reproductive traits in tomato (*Solanum lycopersicum*). *Molecular Breeding* 37 (5): 58.

Ye, C., Argayoso, M.A., Redoña, E.D. et al. (2012). Mapping QTL for heat tolerance at flowering stage in rice using SNP markers. *Plant Breeding* 131 (1): 33–41.

You, F.M., Huo, N., Deal, K.R. et al. (2011). Annotation-based genome-wide SNP discovery in the large and complex *Aegilops tauschii* genome using next-generation sequencing without a reference genome sequence. *BMC Genomics* 12: 59.

Zhou, J., Wang, J., Li, X. et al. (2014). H_2O_2 mediates the crosstalk of brassinosteroid and abscisic acid in tomato responses to heat and oxidative stresses. *Journal of Experimental Botany* 65 (15): 4371–4438.

Zhu, S., Huang, R., Wai, H.P. et al. (2017). Mapping quantitative trait loci for heat tolerance at the booting stage using chromosomal segment substitution lines in rice. *Physiology and Molecular Biology of Plants* 23 (4): 817–825.

5

Alteration in Carbohydrate Metabolism Modulates Thermotolerance of Plant under Heat Stress

Roseline Xalxo[1], *Bhumika Yadu*[1], *Jipsi Chandra*[1], *Vibhuti Chandrakar*[1], *and S. Keshavkant*[1,2]

[1] *School of Studies in Biotechnology, Pt. Ravishankar Shukla University, Raipur, India*
[2] *National Center for Natural Resources, Pt. Ravishankar Shukla University, Raipur, India*

5.1 Introduction

Amid the ever-changing components of the natural environment, the persistently intensifying ambient temperature is considered one of the most detrimental stresses. Heat stress (HS) is often explicated as the elevation in temperature beyond a threshold level, for a period of time, which is sufficient to cause irreversible damage to growth and development processes of flora. Generally, a transient increase in temperature, usually 10–15 °C above ambient, is considered as heat shock or HS (Wahid et al. 2007). Future change in global climatic condition, with a forecast of 1.5–5.8 °C amplification in temperature, leading to global warming, threatens agricultural production (Hemantaranjan et al. 2014). The existing hypothesis about climate change worldwide is that most agricultural land will confront increased environmental fluctuation. Climatological extremes, including very high temperatures, are predicted to have a general negative effect on plant growth and development, leading to a catastrophic loss of crop productivity and quality globally (Yuan et al. 2017). Though HS and other abiotic stresses are limiting factors for crop production, crop yield is often at the

Heat Stress Tolerance in Plants: Physiological, Molecular and Genetic Perspectives, First Edition.
Edited by Shabir Hussain Wani and Vinay Kumar.

mercy of arbitrary environmental variations (Hemantaranjan et al. 2014). Plants, in particular, as sessile organisms, cannot move to more favorable environments. As a result, HS causes an array of amendments at morphological, anatomical, physiological, and biochemical levels, which affects plant growth, development, and even phenology and may lead to drastic decline in crop productivity (Wahid et al. 2007; Hasanuzzaman et al. 2013). At high temperature, cellular injury or cell death might occur within minutes of exposure, which may lead to a disastrous collapse of cellular organization (Wahid et al. 2007).

Elevated temperature imposes damage in all aspects of plant processes like growth, germination, development, reproduction, yield, etc. (Hasanuzzaman et al. 2013). High-temperature stress differentially disturbs the stability of membranes, proteins, and cytoskeleton structure and modulates the efficiency of enzymatic reactions of the cells which obstructs the major physiological processes and creates metabolic imbalance (Suzuki et al. 2012). Elevated temperature also alters the activities of carbon metabolism enzymes, and the accumulation of starch and sucrose biosynthesis by altering specific genes of carbohydrate metabolism (Ruan et al. 2010). At the molecular level, HS modulates the gene expression and accumulation of transcripts, thereby leading to the syntheses of stress-related proteins as a tolerance strategy (Iba 2002). One of the principal consequences of HS is the excess production of reactive oxygen species (ROS), which leads to oxidative stress condition inside the affected cells (Hasanuzzaman et al. 2013; Hemantaranjan et al. 2014; Wang et al. 2018). Plants incessantly struggle for survival under various environmental conditions, including HS. To some extent, these are able to tolerate HS by physical changes within the body, and frequently by generating signals for varying metabolism. HS-induced ROS accumulation is counteracted by enzymatic antioxidant system that includes a variety of scavengers such as superoxide dismutase, ascorbate peroxidase, guaiacol peroxidase, glutathione-S-transferase, catalase, etc., and nonenzymatic low-molecular-weight metabolites like ascorbic acid, proline, glutathione, α-tocopherol, carotenoids, flavonoids, etc., which are able to regulate proteins and various cellular structures, maintain cell turgor pressure by osmotic adjustment, and alter the antioxidant system to re-establish the cellular redox balance and homeostasis (Janska et al. 2010). At the molecular level, HS causes modifications in the gene expression of osmoprotectants, detoxifying enzymes, transporters, and regulatory proteins which are involved in direct defense from HS (Krasensky and Jonak 2012). In various stress conditions, including HS, modifications in physiological and biochemical processes of plants by change in gene expression progressively leads to the development of heat tolerance in the form of acclimation or, in the ideal case, to adaptation (Hasanuzzaman et al. 2010). Recently, exogenous applications of a few of the protectants in the form of osmoregulators, phytohormones, signaling molecules, trace elements, and nutrients have been found to be effective in mitigating HS-induced oxidative damages in plants (Hasanuzzaman et al. 2010).

The development of novel crop varieties which shows tolerance toward HS is a major challenge for plant scientists (Zhang et al. 2006). Attempts to develop heat-tolerant cultivars via conventional plant breeding methods have been successful (Camejo et al. 2006). Recently, though, various advanced techniques of molecular breeding and genetic engineering have provided additional tools, which could be employed to develop heat-tolerant crops, and to fight against universal environment adversary.

This chapter considers plant responses due to the alteration in carbohydrate metabolism under HS at the morphological, anatomical, physiological, biochemical, and genetic levels. This chapter also underlines the mechanisms involved in the thermotolerance of plants, along with strategies adopted for genetic improvement in crops for improving thermotolerance ability.

5.1.1 Heat Stress and Thermotolerance

Of the major forms of abiotic stresses, plants are exposed to HS in nature, which has an independent mode of action on the morphology, physiology, and metabolism of plant cells. The susceptibility to extreme temperature in different plants varies considerably with the stage of its development, species and genotype along with abundant inter- and intraspecific variations (Barnabás et al. 2008; Sakata and Higashitani 2008).

5.1.1.1 Morphological Alterations

Various morphological alterations have been noticed under HS that includes pre- and post-harvest damage, including scorching of leaves and twigs, damage to leaf tips and margins, reduced leaf area, rolling and drying of leaves, necrosis, sunburns on leaves, branches and stem, leaf senescence and abscission, shoot and root inhibition, fruit discoloration and damage, elongation of hypocotyls and petioles, which consequently lead to reduced yield (Hua 2009; Tian et al. 2009; Djanaguiraman et al. 2010). One major deleterious effect of HS on shoot growth is a severe reduction in the first internode length, thereby causing premature death. For example, *Saccharum officinarum* plants grown under high temperature exhibited smaller internodes, increased tillering, early senescence, and reduced total biomass (Ebrahim et al. 1998). In some plant species elevated temperature causes significant elongation of stems, extended leaves (hyponasty), and reduced number of tillers (Djanaguiraman et al. 2010).

5.1.1.2 Anatomical Alterations

At the whole plant level, HS causes reduced cell size, closure of stomata and curtailed water loss, increased stomatal and trichomatous densities, and greater xylem vessel tissues in both roots and shoots (Anon et al. 2004). At the subcellular level, major reorganization of chloroplast occurs, thereby reducing photosynthesis (Karim et al. 1997). Significant increase in the thickness of palisade layers I and II, spongy layer and lower epidermis, has been observed in heat-stressed *Glycine max* (Djanaguiraman et al. 2011). It also decreases stomatal diameter, damages plasma membranes, chloroplasts membranes, and mitochondrial membranes, and decreases the number and size of chloroplasts and mitochondria. On inception of extreme temperature, *Zygophyllum qatarense* produced polymorphic leaves and showed reduced transpirational loss (Sayed 1996). Zhang et al. (2005) report that, under HS, chloroplasts become round in shape, permeability of plasma membrane increases, stroma lamellae swollen, while cristae are disrupted. Such modifications result in antenna-depleted photosystem II and hence reduced photosynthesis and respiratory activities. The cumulative effects of all these changes may lead to poor plant growth and yield responses under HS.

5.1.1.3 Physiological and Biochemical Modifications

Growth Among different phases of plant growth, the foremost stage that is severely affected by various abiotic stresses, including HS, is germination. Diminished germination, seedling emergence, poor seedling vigor, abnormal seedlings, and shortened radicle and plumule growth are major impacts caused by high temperature stress (Kumar et al. 2013; Piramila et al. 2012). Cheng et al. (2009) observed a reduced rate of germination and seedling emergence and enhanced cell death in heat-stressed *Triticum aestivum*. This suppression in plant growth might be the result of altered nitrogen assimilation and carbohydrate metabolism (Yuan et al. 2017). In addition, by disturbing the organization of cell membranes, microtubules, and the cytoskeleton, and altering cell differentiation, elongation, and expansion, HS ultimately hampers the growth and development processes of plants (Potters et al. 2008).

Water Relations Plant water status is the most vital variable under changing ambient temperature (Mazorra et al. 2002). Generally, plants tend to maintain constant water status inside cells regardless of the surrounding temperature when moisture is adequate; on the other hand, high temperature stress severely impairs this tendency when water is limited (Machado and Paulsen 2001). In fields, high temperature is frequently associated with reduced water availability. High temperatures, often mutually with intensive radiation, enhance transpiration via increased stomatal conductance and thus may induce temporal water deficiency in plants even though cells are well watered during the day (Simoes-Araujo et al. 2003). The expeditious transpiration under high temperature causes a noteworthy decline in the water potential of leaves (Anon et al. 2004). Morales et al. (2003) and Wahid and Close (2007) recorded adverse impacts of HS on root hydraulic conductance and leaf water potentials of *Solanum lycopersicum* and *S. officinarum*, respectively, even when soil water supply and relative humidity were favorable. Likewise, high surrounding temperature reduces the water status of flower buds and leaves, which in turn leads to improper pollen development (Tsukaguchi et al. 2003). Suzuki et al. (2001) suggest that together with limited assimilate supply the deterioration of water status in flower buds causes a drastic decline in pollen fertility, and thereby yield. Moreover, elevated temperature appears to cause more water loss during the day than at night.

Photosynthesis and Stomatal Conductance HS induces changes in various vital physiological processes of plants, and shortens the lifecycle and reduces productivity. Photosynthesis is one of the most heat sensitive physiological processes (Crafts-Brandner and Salvucci 2002). Structural modifications in the plant cells exposed to elevated temperature cause alterations in the energy allocation to the photosystems (Wahid and Shabbir 2005; Allakhverdiev et al. 2008). High temperature stress has a greater influence on the photosynthetic capacity, especially of C_3 plants rather than C_4 plants (Yang et al. 2006). Decreased photosynthetic efficiency may be due to the inhibition of enzymes of the Calvin cycle, including ribulose-1,5-bisphosphate carboxylase/oxygenase (Rubisco) (Morales et al. 2003). Decline in chlorophyll content due to increased chlorophyllase activity and decreased amount of photosynthetic pigments also leads to reduced photosynthesis and respiratory activity (Todorov et al. 2003). Stomatal closure due to HS is another cause for impaired photosynthesis that affects the intercellular CO_2 concentration (Ashraf and Hafeez 2004). The reduction of soluble proteins, Rubisco-binding proteins, large and small subunits of Rubisco, etc., can also hamper photosynthesis (Sumesh et al. 2008). High-temperature stress also adversely affects syntheses of starch and sucrose by reducing activities of sucrose phosphate synthase (SPS), ADP-glucose pyrophosphorylase, and invertase (INV) (Djanaguiraman et al. 2009).

5.1.1.4 Cell Membrane Integrity

Continuous functioning of cellular membranes under various abiotic stresses is vital for essential processes such as photosynthesis, respiration, etc. Any change in ambient temperature is immediately recognized by plants with a complicated set of sensors located in various cellular compartments. Elevated temperature accelerates the kinetic energy and movement of biomolecules across the membranes resulting in a loosening of chemical bonds within the molecules of membranes.

Such a loosening of chemical bonds leads to disintegration of membrane lipids and an increase in membrane fluidity by either protein denaturation or an increase in unsaturated fatty acids caused by extreme temperature (Savchenko et al. 2002). The integrity and functioning of biological membranes are sensitive to high temperature, as it modifies the tertiary and quaternary structures of proteins. Such modifications lead to increased membrane permeability, which increases the loss of cellular electrolytes. Increased membrane fluidity also leads to the activation of lipid-based signaling cascades, increased Ca^{2+} influx, and the reorganization of the cytoskeleton. Such signaling process results in production of osmolytes and various antioxidants in response to high temperatures (Bita and Gerats 2013). For instance, the *CNGC2* gene of *Arabidopsis thaliana* encodes a component of the membrane cyclic nucleotide gated Ca^{2+} channels that functions as the primary thermosensor of cells. These plasma membrane channels respond to increased ambient temperature by activating an optimal heat shock response (Saidi et al. 2009). The increased solute leakage, as an indication of decreased cell membrane integrity, has long been used as an indirect measure of HS tolerance. High-temperature stress also uncouples electron transport from photophosphorylation, most probably owing to the increased permeability of thylakoid membranes (Schrader et al. 2004). Rate of electrolyte leakage has been shown to be influenced by various factors, including plant/tissue age, organ, developmental stage, growing season, degree of hardening, and species. It has been suggested that membrane fluidity is linked to the degree of unsaturation of membrane lipids (Los and Murata 2004). An increase in saturated fatty acids in mature leaves increases the melting temperature of plasma membranes and thus reduces heat tolerance of the plants.

5.2 Carbohydrate as Protectives Molecules

5.2.1 Osmolyte

The accumulation of osmoprotectants is a vital adaptive mechanism in plants when subjected to abiotic stresses including high temperature, as primary metabolites are directly indulged in osmotic adjustment. For example, the accumulation of soluble sugars is fundamental for the regulation of osmotic activities. This protects various important cellular structures from elevated temperature stress by maintaining the cell–water balance and membrane stability and by buffering the cellular redox potential (Nishizawa et al. 2008). Apart from acting as osmoprotectants, these molecules play the role of antioxidants, as observed in the leaves of *A. thaliana*. Sucrose has been reported as the most prevalent carbohydrate in leaves and can scavenge hydroxyl radical *in vitro* (Smirnoff and Cumbes 1989). Another carbohydrate mannitol is a widely distributed sugar alcohol in organisms from bacteria to algae, fungi and higher plants, and functions as an osmoprotectant (Bieleski 1982). Wahid and Close (2007) report that the accumulation of soluble sugars has great implications for heat tolerance in *S. officinarum*. Disruption in fruit setting in *S. lycopersicum* due to the disruption in metabolism of sugar and transport of proline during pollen development under HS was documented by Sato et al. (2006).

5.2.2 Thermoprotectant

During stress conditions including high temperature, carbohydrate metabolism plays a significant role in the tolerance mechanism as it is directly linked to photosynthetic performance of the plants. Kaplan et al. (2004) detected significant change in amino acids and carbohydrate metabolisms of *A. thaliana* when exposed to HS. Carbohydrates that are found to accumulate during HS include maltose, galactinol, sucrose, myo-inositol, and raffinose, together with cell wall precursors, mainly monosaccharides. Metabolic profiling revealed that plants subjected to a combination of drought and HS accumulated sugars like glucose, fructose, sucrose, trehalose, and maltose, and helped to maintain cell turgor, stabilizing cell membranes and preventing protein degradation (Huang et al. 2013). Also, during adverse conditions, plants use starch and fructan as energy sources and substitutes of glucose (Kaplan et al. 2004). Indeed, under circumstances like water deficit, concentrations of soluble carbohydrates such as fructose and glucose increased, while sucrose was transported to the root tips to promote growth (Nishizawa et al. 2008). Likewise, over accumulation of trehalose by transgenic plant also showed an enhanced tolerance to HS. In addition, a few oligosaccharides, namely raffinose and stachyose with no energetic role, also accumulate in response to a range of abiotic stresses, and have been shown to be involved in reducing oxidative damage and ROS scavenging (Kaplan et al. 2004).

5.3 Carbohydrates as Signaling Molecules

In plants, carbohydrates as well as being the major sources of carbon and energy which function as substrates for growth and development also modulate the expression of genes, thus playing a role as signaling molecules in many physiological processes (Granot et al. 2013). Carbohydrate metabolism has been reported as the crucial metabolic and transcriptional component that dictates plant's sensitivity to abiotic stresses (Li et al. 2015). Therefore, alterations in sugar metabolism due to HS adversely affect sugar-sensing systems.

5.3.1 Reproductive Cell Development

Among all the growth and developmental stages of plants, the reproductive tissues are the most sensitive and are susceptible to HS, and a small increase/decrease in temperature during flowering can lead to the loss of an entire grain crop cycle (Lobell et al. 2013). During the reproductive stage, high temperature, even of a short duration, causes decrease in buds, and flower abortion (Sato et al. 2006). Plants susceptible to HS may not produce flowers, or flowers may not produce fruits or seeds (Foolad 2005; Maheswari et al. 2012). Hence, reasons for increased sterility under abiotic stress including high temperature are impaired gametogenesis, altered pollen germination and tube growth, reduced ovule viability, anomaly in stigmatic and styles positions, altered number of pollen grains retained by stigma, disrupted fertilization and post-fertilization processes, and hindrance in growth of endosperm, pro-embryo and fertilized embryo (Foolad 2005). Both female and male gametophytes are sensitive to HS, and the response of plants to HS varies with genotype; however, the male gametophyte is particularly more sensitive at all developmental phases, while the pistil and the female gametophyte are considered more tolerant (Hedhly 2011). Saini (1997) reports that stress-induced disturbance in carbohydrate metabolism and distribution within anthers adversely affect the development of the male gametophyte. These defective pollen grains were impotent to accumulate starch, which is the principal component of fertile grass pollen. Hindrance of sugar metabolism is also reported to be associated with failure of viable pollen grain production under HS conditions in *S. lycopersicum.* High-temperature exposure at heading stage (when the panicle is fully visible) drastically reduces dehiscence of anther and pollen fertility rate, leading to a decrease in the number of pollens on the stigma, which ultimately leads to reduced fertilization and subsequent spikelet fertility and seed sterility in *Oryza sativa* (Cao et al. 2008). Elevated temperature often causes excessive production of ethylene, which results in male sterility of *O. sativa* pollens via hampering key enzymes of sugar starch metabolism, which enfeebles sink strength and limits grain filling and eventually leads to the production of sterile grains. Overall, based on the studies, it seems that in most of the plants, reproductive processes are remarkably affected by HS, which eventually affects fertilization and post-fertilization processes leading to reduced crop production.

5.3.2 Seed Development

High temperature throughout the seed development may culminate in reduced germination and loss of vigor, leading to diminished seedling emergence and establishment, as has been observed for various crop plants (Akman 2009; Ren et al. 2009). Both grain weight and number appears to be influenced by high-temperature stress, and are directly proportional with amplified temperature during flowering and grain filling (Mohammed and Tarpley 2010). Reduction in starch and protein qualities and total oil yield in several crop species have been associated with HS during seed development (Banowetz et al. 1999). In *T. aestivum*, high-temperature stress during reproductive development hastens the decrease in photosynthesis and leaf area, reduced shoot and grain mass, and also decreased weight and sugar contents of kernels along with reduction in water use efficiency (Shah and Paulsen 2003).

5.3.3 Seed Germination and Yield Loss

High-temperature stress is raising alarm worldwide, regarding crop productivity and food security (Hasanuzzaman et al. 2013). The effect of elevated temperatures is so frightening that even a minute (1.5 °C) rise in temperature has noteworthy negative consequences on seed germination and crop yields (Warland et al. 2006). Germination process is regulated by the enzyme trehalose-6-phosphate (T6P) via regulating the glycolytic flux of carbohydrate metabolism (Delorge et al. 2014). During HS the germination process of seed is disturbed, owing to impaired function of enzymes that are associated with starch breakdown (Essemine et al. 2010). High-temperature-induced reduction in productivity has been documented by many scientists in various cultivated crops, including cereals (e.g. *O. sativa, T. aestivum, Hordeum vulgare, Sorghum bicolor, Zea mays*), pulse (*Cicer arietinum, Vigna unguiculata*), and oil-yielding crops like *Brassica nigra* and *Brassica napus* (Zhang et al. 2013; Maheswari et al. 2012; Cao et al. 2008; Warland et al. 2006; Kalra et al. 2008). Decrease in crop productivity due to HS is chiefly associated with decreased assimilatory capacity (Sinsawat et al. 2004), which is due to diminished photosynthetic rate and altered membrane stability (Zhang et al. 2006). High-temperature stress also affects the performance and quality of crop. Hogy et al. (2013) observe that grain quality characteristics are significantly changed under HS in *H. vulgare*, which may be related to reduction in nonstructural carbohydrates like fructose, starch, and raffinose. Gunawardhana and de Silva (2011) observed damage in pod quality parameters like fiber content and fragmentation of Ca^{2+} pectate in *Abelmoschus esculentus*.

5.4 Adverse Impacts of Heat Stress

5.4.1 Photosynthesis

HS imposes negative impacts like reduced water potential and area and premature senescence in leaves, which all have an adverse effect on the total photosynthetic performance of plants (Greer and Weedon 2012). Under prolonged HS, depletion of carbohydrate reserves and plant starvation have also been observed (Djanaguiraman et al. 2009). Decline in the average rate of photosynthetic activity due to HS was observed in many of the temperature-sensitive genotypes like *O. sativa*, *Vitis vinifera*, etc. (Hurkman et al. 2009; Greer and Weedon 2012), and this diminution was attributed to altered carbon assimilation owing to stomatal closure and/or chlorophyll breakdown.

5.4.1.1 Altered Carbon Assimilation

Stomatal closure due to HS is another cause of impaired photosynthesis that affects the intercellular concentration of CO_2 (Ashraf and Hafeez 2004). Elevated temperature reduces photosynthesis, while enhances CO_2 transfer between intercellular spaces and carboxylation sites. In various plants, stomatal conductance and net photosynthesis are inhibited by high-temperature stress due to the decrease in the activation state of Rubisco (Crafts-Brandner and Salvucci 2002; Morales et al. 2003). Low affinity of Rubisco for CO_2 also affects the rate of photosynthesis under high temperature. For instance, in *Z. mays*, photosynthesis was restrained at leaf temperatures above 38 °C, and this inhibition was much more severe when the temperature was increased further, although this inhibition was independent of stomatal response to high-temperature stress (Crafts-Brandner and Salvucci 2002).

5.4.1.2 Chlorophyll Breakdown

HS causes reduction in chlorophyll content thereby impairing the photosynthetic pathway. Initial effects of HS encompass structural modifications in protein complexes of chloroplasts and reduced enzyme activity. In chloroplast, carbon metabolism of the stroma and photochemical reactions in thylakoid lamellae are considered the primary sites of injury due to HS (Wang et al. 2009). Particular effects of HS on photosynthetic apparatus include membrane damage, impaired structural organization of thylakoids, swelling of grana stacks, and loss of grana stacking (Ashraf and Hafeez 2004). Such structural modification in the plant cells exposed to elevated temperatures leads to alteration in the energy allocation to the photosystems (Wahid and Shabbir 2005; Allakhverdiev et al. 2008). Decline in chlorophyll content due to increased activity of chlorophyllase, and decreased amount of photosynthetic pigments, also leads to reduction in the plant's photosynthetic and respiratory activities (Todorov et al. 2003).

5.4.2 Major and Minor Carbohydrate Metabolism

The genes and enzymes involved in energy metabolism like glycolysis, photosynthesis, and ATP biosynthesis (ATPase, ferredoxin, cytochrome C) are highly altered after HS. MapMAN pathway after short- and long-term enrichment analyses showed six pathways, including both major (glycolysis, N-metabolism) and minor (amino acid metabolism, cofactor, and vitamin metabolism) carbohydrate metabolism. Short-term heat treatment specifically influences glycolysis, nucleotide metabolism, photosynthesis, oxidative phosphorylation, tetrapyrrole synthesis, fermentation, etc., while long-term heat treatment alters the pathways of metal handling, redox homeostasis, polyamine metabolism, etc.

5.4.3 Expression of Regulatory Genes

Abiotic stresses, like salinity, drought, and HS, have a strong impact on gene expression patterns. Expressions of sugar-responsive genes are either boosted or suppressed by an abundance or scarcity of carbohydrate. The expression of genes varies depending on the type of sugar, though metabolic flux may be more important than actual levels of carbon resources. In heat-sensitive *O. sativa* plants, expression of a carbon-starved anther gene was reported to be enhanced under HS, which encodes sugar transporter (*MST8*) and cell wall INV, and was identified as marker of heat-tolerant plants (Li et al. 2015). Different techniques, namely DNA microarray, differential hybridization, and reverse transcription-PCR were undertaken to monitor the level of gene expression and examine the regulation of starch synthesizing different genes like granule-bound starch synthase-I and branching enzymes, which are reported to be associated with grain filling in *O. sativa* (Jha et al. 2014).

A number of sugar-modulated genes have direct or indirect roles in the syntheses of sugar-metabolizing enzymes, which suggests that their altered expressions may have adaptive values. Genes encoding proline and nitrate transporters were unregulated during HS, implied that plants had tried to maintain the steady photosynthesis and minimize the damage caused by high temperature. During HS, microRNAs (miRNAs) regulates the expression of metabolic pathways and stress-associated genes, which can modulate the thermotolerance by manipulating the activities of key enzymes.

5.4.4 Enzyme Activity

Several gene-encoding enzymes involved in cellular metabolism are differentially expressed upon exposure to abiotic stresses, and modulation of some of the stress-related transcription factors (TFs) have been shown to induce changes in stress associated metabolites. High-temperature stress also adversely affects syntheses of sugars by reducing activities of carbohydrate-metabolizing enzymes. However, during HS, levels of soluble sugars were observed to be not affected by alteration in the surrounding temperature, when the enzymes of sugar metabolism were functioning under coordinated control.

- Sucrose synthase (SS; EC 2.4.1.13)
  Sucrose synthase (SS) is considered a main catalytic enzyme for sucrose cleavage in many tissues, and its activity can be used as a biochemical marker for measuring sink strength in plants (Xu et al. 1989). The activity of SS may be reduced in heat-sensitive genotypes, and also affects the overall yield, by altering the carbohydrate partitioning from roots to shoots, whenever HS is experienced (Wolf et al. 1990). High-temperature stress reduces the soluble sucrose synthase (SSS) activity, which is required for amylopectin synthesis, an important constituent of starch in the endosperm (Delvalle et al. 2005). HS during endosperm development is an important yield-limiting factor, as the rate of assimilation, translocation, and duration of grain filling are directly influenced by alterations in ambient temperature (Liu et al. 2011). Altered SSS activity under HS also disintegrates and distorts starch granule synthesis (Kumar et al. 2013).

- Soluble acid invertase (SAINV, EC 3.2.1.26)
  As sugar acts as signaling molecule during abiotic stresses, increased levels of total soluble sugar may be linked to increased INV activity, which is an adaptive mechanism for exerting protective roles under HS (Kaur et al. 2009). While, on the other side, HS can block the sucrose import and reduce the activity of INV, deplete starch reserves, and cause a dramatic reduction of hexoses in the reproductive organs, consequently causing abortion (Ruan 2014).
- Sucrose phosphate synthase (SPS, EC 2.4.1.14)
  Sucrose is the main form of translocated sugar in most plants. Its synthesis is catalyzed by SPS and sucrose-6-phosphate, while degradation pathway involves SS or INV (Preiss 1982). During HS, a significant alteration in the activity of SPS has been observed in heat-sensitive genotypes, which results in the reduced synthesis of sucrose (Zhou et al. 2016). Its reduction results in the impairment of pollen function, fertilization, and fruit setting (Parrotta et al. 2016). SPS is the major control point for the partitioning of photosynthate between sucrose and starch, which was further confirmed by the overexpression of *Z. mays* SPS in *S. lycopersicum* leaves to augment the foliar sucrose concentration but declined the starch content (Galtier et al. 1993). Wallwork et al. (1998) found that *H. vulgare* grains from heat-treated plants accumulated less starch than grains from control plants, owing to a reduced conversion of sucrose to starch. Sheoran and Saini (1996) suggest that stress could inhibit starch deposition in pollen, either by decreasing the availability of assimilates or by impairing the activities of enzymes involved in starch biosynthesis. This has led to the suggestion that the accumulation of sugars could be the result of disturbances in carbohydrate metabolism or of an inhibition of sugar utilization (Saini 1997).
- ADP-glucose pyrophosphorylase (AGPase, EC 2.7.7.27)
  ADP-glucose pyrophosphorylase has also been shown to be an important enzyme that is associated with starch synthesis. It is reported to be associated with growth and maturation of tubers in *S. tuberosum* as its activity declines after the maturation and detachment of tubers (Geigenberger et al. 1994). Transgenic mutants with tuber-specific expression

of an AGPase clone insensitive to allosteric regulation accumulate more starch in tubers (Stark et al. 1992).

- Ribulose-1,5-bisphosphate carboxylase/oxygenase (Rubisco, EC 4.1.1.39)
  During dark reactions of photosynthesis, activation of Rubisco in the Calvin cycle has been determined as a critical step, being inhibited at 35–40 °C, which results in decreased net CO_2 assimilation and the production of carbohydrates (Dubey 2005). High temperature influences the photosynthetic capacity of C_3 plants more strongly than in C_4. It alters energy distribution and changes the activities of carbon metabolizing enzymes, particularly Rubisco, thereby altering the rate of RuBP regeneration by the disruption of electron transport and inactivation of the oxygen evolving enzymes of photosystem II (Salvucci and Crafts-Brandner 2004). Heat shock reduces the amount of photosynthetic pigments (Todorov et al. 2003), soluble proteins, Rubisco-binding proteins, and large and small subunits of Rubisco under darkness but increases them under light conditions, indicating their roles as chaperones and heat shock proteins (HSPs) (Kepova et al. 2005).
- Galactinol synthase (GolS, EC 2.4.1.123)
  Biosynthesis of raffinose family oligosaccharides (RFOs), such as raffinose, stachyose and verbascose, is initiated by the formation of galactinol from myo-inositol and UDP (uridine diphosphate) galactose by GolS. Sequential addition of galactose units, provided by galactinol, to sucrose leads to the formation of raffinose and higher-order RFO (Peterbauer and Richter 2001). *A. thaliana* plants overexpressing *Arabidopsis* $GolS_1$ or $GolS_2$ accumulated high levels of galactinol and raffinose, respectively, and were more tolerant to abiotic stresses (Nishizawa et al. 2008). By contrast, impaired raffinose accumulation showed an obvious increase in sensitivity to heat or freezing stress (Zuther et al. 2004).
- Trehalose-6-phosphate synthase (TPS, EC 2.4.1.15)
  In plants, trehalose is synthesized in a two-step process (Paul et al. 2008). Trehalose-6-phosphate synthase (TPS) generates T6P from UDP-glucose and glucose-6-phosphate followed by dephosphorylation to trehalose by trehalose-6-phosphate phosphatase (TPP). Transgenic expression of trehalose

biosynthesis genes showed that enhanced trehalose metabolism can positively regulate tolerance to abiotic stress, even though only a limited increase in trehalose content could be observed, excluding a direct protective role of trehalose in these plants. Consistently, loss of TPS5 function, a TPS with a TPP domain, has been found to reduce the basal thermotolerance of *A. thaliana* (Suzuki et al. 2008).

5.5 Mechanisms Involved in Thermotolerance

In plants, during photosynthesis, sugars are produced that serve as substrates for energy metabolism and precursors of polysaccharides like starch and cellulose. Also, sugars act as a signaling molecule and plays a crucial role in seed germination, growth, and protection during exposure to abiotic stresses. As a primary messenger, sugars regulate signals that control the expression of various genes involved in carbohydrate metabolism under abiotic stresses. Accumulation of sugars during stressed conditions may contribute to membrane protection, osmotic balance, free radical scavenging, maintenance of turgor pressure, and source of food reserves to allow plants to recover from adverse conditions (Chandrakar et al. 2018). Its accumulation also prevents dehydration and oxidation of cell membranes, and proteins under HS. Exposure to mild heat leads to the accumulation of sugar, which may be due to the inhibition in the activities of SS or INV, while at extreme high temperatures sugar content is reduced, which might be related to stress-induced reduction in photosynthesis and the subsequent scarcity of photoassimilates (Harsh et al. 2016). Transgenic studies offer new and challenging opportunities for a better understanding of the mechanisms implicated in the improvement of stress tolerance by inserting a gene, involved in protection from any stressor, into agriculturally important crop plants.

5.5.1 Glucose and Heat-stress Tolerance

Sugars like glucose are signaling molecules, and serve as one of the important parts of living cell structures and protect the plants under stressed conditions (Huang et al. 2015; Yadu et al. 2016). Among carbohydrates, accumulations of glucose and fructose are more closely related to heat tolerance in plants (Liu and Huang 2000). Under abiotic stresses such as heat, drought, salt, etc., pretreatment with glucose decreases the oxidative stress by reducing the extent of lipid peroxidation, regulating photosynthesis and activities of antioxidant enzymes (Huang et al. 2013; Chandrakar and Keshavkant 2018). Heat-tolerant species tend to minimize the toxic effects of stress by maintaining a lower rate of respiration (Liu and Huang 2000). Marias et al. (2017) report that in expanding leaves of *Coffea arabica* there were more accumulations of glucose and fructose under HS because, in the cell walls of these leaves, glucose and fructose were the crucial building blocks for lignin, cellulose, etc. (structural carbohydrates), essentially required for the expanding leaves. In plants, HS results in the overaccumulation of ROS and malondialdehyde (MDA), and disorganized the chloroplast structure; however, the exogenous application of glucose reduces the percentage of abnormal chloroplasts and prevents the ultra-structure of chloroplasts, which coincides with less ROS and MDA accumulation. Also, glucose pre-treatment up-regulates the transcript levels of antioxidant enzymes thereby increasing HS tolerance (Huang et al. 2015).

5.5.2 Sucrose and Heat-stress Tolerance

Sucrose is a nonreducing disaccharide chiefly synthesized in the cytoplasm of mesophyll cells of mature leaves. It is present in both source and sink organs, and is utilized as a source of carbon which depends on its cleavage by two enzymes, namely INV and SS into hexoses. The expressions of genes encoding INV and SS are highly affected by HS. Other enzymes such as SPS and SAINV work in combination with SS during hydrolysis of sucrose to provide hexoses for various structural and functional requirements such as energy generation from macromolecules like starch (Chandrakar et al. 2016; Yuan et al. 2017).

Transcriptional studies revealed that the gene expression of SS increases to an extent, owing to the increased demand of sucrose and its metabolites to meet out elevated energy requirements of the cell during mild HS. But, at extreme temperatures, this enzyme is denatured or inactivated, thereby affecting the overall carbon metabolism. Also, under HS, higher SPS transcript levels might contribute to creating carbon gradient flux from starch to sucrose, hence results in sucrose accrual. Moreover, higher expressions of SS and INV degrade starch into products that are exported to cytoplasm and are again converted into sucrose via SPS. Thus, sucrose content increases, which might trigger rapid recycling of stored carbon from a source to a sink (Yuan et al. 2017). Lafta and Lorenzen (1995) report that, at high temperatures, enhancement in the expression of SPS genes causes increase in the synthesis of sucrose that enhances tolerance of plants under mild HS.

5.5.3 Fructan and Heat-stress Tolerance

Fructan is a branched polyfructose molecule that is synthesized by many plants in the vacuole by fructosyltransferases, which transfer fructose from sucrose to the growing fructan chain (Livingston et al. 2009). Under stressed conditions, tonoplast-derived vesicles carry fructan from the vacuole to the apoplast. The accrual of fructan may play a role in adaptation to HS, owing to their high water solubility and ability to stabilize membranes, and may indirectly contribute to osmotic adjustment by the release of hexose sugars (Valluru and Van den Ende 2008). Their accumulation along with starch as storage stuff can be utilized by plants during starvation or enhanced demand of energy (Krasensky and Jonak 2012). In terms of quantity, the accrual of fructan is the dominant among all the carbohydrate in plants under HS (Liu and Huang 2000). Fructan might also act as an endogenous signal that mediates phloem-mobile stress signals. Binding efficiency of fructan between the head groups of lipid leads to slowing of their molecular motion and substantial depletion of the outflow of water from the dry membrane under HS (Valluru and Van den Ende 2008). In response to HS, the accumulations of fructan also prevents plasmolysis by increasing the osmotic pressure of the cells (Livingston et al. 2009). Fructan minimizes the toxic effects of HS by membrane protection via protecting the liposomes from leakage and fusion.

5.5.4 Trehalose and Heat-stress Tolerance

Trehalose, also known as tremalose and mycose, is a nonreducing disaccharide that is greatly accumulated and provides tolerance to plants experiencing abiotic stress, including heat (Gupta and Kaur 2005). Naturally, it is present in trace amounts in plants but is highly accumulated under abiotic stresses (Kempa et al. 2008). Trehalose can act as an osmolyte and plays an important role in maintaining cell volume and fluid balance. Also, it has been implicated in the stabilization of membranes and proteins under stressed conditions (Paul et al. 2008). In higher plants, the first step of the trehalose biosynthesis pathway is catalyzed from UDP-glucose and glucose-6-phosphate by TPS to produce T6P in which dephosphorylation occurs by TPP, and trehalose is synthesized (Paul et al. 2008). Transgenic studies reveal that up-regulation in the expression of trehalose-synthesizing genes enhances the trehalose metabolism that positively regulates tolerance to abiotic stresses.

Exogenous application of trehalose ameliorates the toxicity symptoms of HS in *T. aestivum* (Luo et al. 2014). Also, trehalose is involved in the metabolic signaling pathways and plays a role in the regulation of carbohydrate metabolism. HS-induced oxidative conditions lead to reduction/inhibition of seed germination. In this condition, accumulation of trehalose emerges as a significant source of energy which can be utilized by the germinating seeds for their recovery. The germination processes are also induced by the precursor of trehalose, namely T6P that plays an important role in the regulation of glycolytic flux of carbohydrate metabolism (Delorge et al. 2014). The reports of Doehlemann et al. (2006) clearly reveal that, owing to the disruption to the *TPS* gene, there is a scarcity of trehalose, which limits the carbon source and increases the sensitivity of germination and growth to high-temperature-induced oxidative stress and loss of viability. Trehalose-mediated stabilization of biological structure and function under HS may be related to three major mechanisms: (i) in dehydrated conditions, trehalose may work as water replacement by the formation of hydrogen bonding with the polar heads of lipids and proteins, thereby stabilizing the native confirmation of biomolecules; (ii) it entraps and concentrates the residual water molecules surrounding the surface of biomolecules, hence can maintain their structure and stability; and (iii) it acts as a vitrifying agent and has the tendency to form less reactive and protective structures which are stable at high temperature. Such protective structures hold the biomolecules and allow them to retain their structures and function (Doehlemann et al. 2006).

5.5.5 Raffinose and Heat-stress Tolerance

Raffinose is an α-galactosyl derivative of sucrose and constitutes carbohydrate reserves of seeds. It accumulates in leaves of plants in response to HS (Kempa et al. 2008). Also, it is a well-known membrane protector and free radical scavenger (Nishizawa et al. 2008). Raffinoses are synthesized in the pathway where first galactinol is produced from myo-inositol and UDP-galactose by galactinol synthase.

HS induces overproduction of ROS, and one of the most reactive ROS is the ˙OH that is formed from H_2O_2 by the Haber–Weiss or Fenton reactions. This ˙OH can potentially react with all the cellular molecules. Because of the absence of a potential enzymatic defensive mechanism for the elimination of this radical, its excess production leads to cell death. Raffinoses has a key property of reacting with ˙OH by hydrogen addition, abstraction, and/or electron transfer. This efficient scavenging of ˙OH in chloroplasts is essential for the proper maintenance of photosynthesis, and is taken care of by raffinoses, partly, under conditions of HS (Nishizawa et al. 2008).

5.6 Genetic Approaches/Strategies for Improving Thermotolerance

5.6.1 Genetically Modified Crop Production

Crops are relentlessly challenged by several types of abiotic stresses, including chilling, drought, salinity, heat, etc., resulting in modifications at the physiological, biochemical, and molecular levels that in turn adversely affect their growth and productivity (Gupta and Kaur 2005; Wahid et al. 2007; Hasanuzzaman et al. 2013). Therefore, to fulfill the demand of a rapidly growing population, it is a big task to increase global food production considerably, but with less exploitation of land, water, and energy, and meanwhile preserving the fertility of the soil and natural resources (Bita and Gerats 2013; Hasanuzzaman et al. 2013; Guo et al. 2016). Nowadays, advancement has been made in production of stress-tolerant crops through conventional selection, breeding, or modern molecular biological approaches. In recent years, efforts have been made on types of genes that regulate and/or are involved in metabolic pathways of improving/producing HS-tolerant crop varieties (Hemantaranjan et al. 2014; Guo et al. 2016).

To develop stress-tolerant varieties, genetic engineering has been widely exploited to insert genes from distant gene pools into the targeted plants. There is an urgent requirement to produce such types of crops having complex traits like abiotic stress resistance and nutritionally balanced as well as amalgamation of multiple traits, as the introduction of single genes for the development of a single trait is not enough (Wahid et al. 2007; Bakhsh and Hussain 2015). For quality foods, genetically modified (GM) crops can perform as a potent complement to the crops developed by strenuous and time-consuming conventional breeding techniques. These GM crops expressing numerous stress-inducible genes have shown increased tolerance against heat, drought, salinity, etc., and also can grow more rapidly compared to traditional/conventional breeding, and the time required for their development being reduced from years to months (Hemantaranjan et al. 2014; Lamaoui et al. 2018). Additional profits of GM crop generation incorporates increased yield, enhanced resistance to insects, pests, diseases, weather conditions, and herbicides (Hemantaranjan et al. 2014). Stress-specific genes utilized to generate tolerant GM crops can either be engaged in signal transduction pathways or can be associated with categories of TFs such as *AP2/EREBP (ERF)*, *bZIP*, *Cys2His2 zinc-finger*, *MYC*, *NAC*, *MYB*, and *WRKY*, which are involved in regulating plant stress responses (Shinozaki et al. 2003; Bakhsh and Hussain 2015; Khan et al. 2015; Obaid et al. 2016). By the means of cis-elements located in the promoter regions of different stress-linked genes, the TFs interact and thus up-regulate the expression of a number of downstream genes, which results in abiotic stress tolerance in crop plants (Biłas et al. 2016), for example the overexpression of the *OsWRKY11* gene using the heat-inducible *HSP101* promoter of transgenic *O. sativa* plants, the NaCl-triggered expression of *SbDREB2A* transcripts in *Salicornia brachiata*, and the overexpression of *CarHSFB2* gene in transgenic *A. thaliana* demonstrate considerable tolerance toward HS (Wu et al. 2009; Gupta et al. 2010; Ma et al. 2016). Thus, TF-based genetic engineering provides a tremendous stratagem for crop improvement as it can serve as an excellent candidate for amending complex traits like HS tolerance in crop plants that are inherently multigenic in nature (Hemantaranjan et al. 2014; Nadeem et al. 2018).

5.6.2 Transgenic Strategies

Following conventional breeding and transgenic strategies, several attempts have been made to detect or increase a plant's tolerance to HS, and these attempts have verified that HS tolerance is a multigenic trait (Wahid et al. 2007; Hasanuzzaman et al. 2013). At different developmental stages of plants, there are various components responsible for tolerance to heat or other abiotic stresses which are regulated by different sets of genes (Wahid et al. 2007). Currently, a great deal of information regarding physiological and metabolic aspects of plant tolerance toward extreme temperature is available. Comparatively, there is limited knowledge about the genetic basis of heat tolerance; however, genetic characterization and/or production of HS-tolerant crop varieties is now possible with the use of traditional plant-breeding methods and recent molecular biological techniques, consisting genetic transformation and molecular marker technology (Nadeem et al. 2018). The genetic relationship between diverse stresses and tolerance mechanisms can be deciphered by quantitative trait locus (QTL) mapping. To recognize specific chromosome segments containing candidate genes for heat tolerance, QTL mapping is the preferential technique used by researchers (Bita and Gerats 2013; Jha et al. 2014; Nadeem et al. 2018).

To improve plants' abiotic stress tolerance, basically there are two common biotechnological approaches/strategies including marker-assisted selection (MAS) and genetic transformation. MAS has been regarded as an effectual approach, owing to the complexity in phenotype selection for tolerance and the general intricacy of abiotic stress tolerance. MAS and genetic transformation requires recognition as genetic markers that can be associated with genes or QTLs affecting individual components or whole plant stress tolerance (Foolad 2005; Bita and Gerats 2013). In several crop species, ample research has been carried out during the last two decades to identify genetic markers associated with different environmental stresses like low temperature, drought, and salinity. However, relatively narrowly defined research has been conducted regarding HS tolerance in different plant species to identify genetic markers (Hasanuzzaman et al. 2013; Nadeem et al. 2018). For instance, in *A. thaliana*, using a set of heat-sensitive mutants four genomic loci (QTLs) impelling its ability to acquire thermotolerance were identified (Hong and Vierling 2000).

During rapid HS, syntheses and accretion of specific proteins like HSPs are ascertained. When plants experience either sudden or regular increase in temperature, the production of these HSPs is elevated significantly. Thus, the most imperative aspect of thermotolerance is the immense production of HSPs (Guo et al. 2016). The up-regulation of some of the heat-inducible genes, which encode HSPs, are very much required for plants' survival under high temperatures (Wahid et al. 2007). Lee et al. (1995) produced transgenic HS-tolerant *A. thaliana* successfully by modulating the TF *AtHSF1* responsible for synthesis of HSP. Another study, done by Sanmiya et al. (2004), showed that the transformed *MT-sHSP* gene exhibited thermotolerance in *Nicotiana tabacum*. Xu et al. (2013) report that in *A. thaliana* overexpression of five *HSP90* genes of *G. max* revealed a decline in lipid peroxidation and increased biomass production under HS. Additionally, overexpression of *OsHSP26* (sHSP) gene of *Escherichia coli* awarded enhanced tolerance to *O. sativa* against HS (Lee et al. 2000). In addition, several environmental stressors, namely drought, salinity, and low temperature, result in modifications in the carbohydrate metabolism, and in order to introduce alteration in the metabolism the sugar signaling pathways interact with the stressors (Gupta and Kaur 2005; Li and Sheen 2016). Under various abiotic stresses, sugar molecules indirectly play a vital role during growth and development of plants by regulating carbohydrate metabolism. A huge number of stress-responsive genes have been documented that are stimulated by sugar moieties, signifying the role of sugars in abiotic stress responses (Gupta and Kaur 2005; Tayade et al. 2018). Suzuki et al. (2005) compared the sugar profiling of transgenic and wild-type plants exposed to HS, and their results reveal that under controlled conditions the relative level of trehalose was higher in *MBF1c*-expressing plants in comparison to the wild-type plants, which was further increased in transgenic plants in response to HS. Miranda et al. (2007) and Suárez et al. (2009) have documented an increase in thermotolerance when transgenic *A. thaliana* and *Medicago sativa* plants expressing the *Saccharomyces cerevisiae* homologues of *TPS* and *TPP* were exposed to HS. Likewise, in double knockout mutants *hsfA1a/hsfA1b* of *A. thaliana*, alterations in the patterns of genes related to sugar metabolism were observed that revealed thermotolerance in this phenotype to some extent (Busch et al. 2005). Likewise, the overexpression of gene *HsfA2* causes enhanced levels of galactinol and raffinose in *A. thaliana* (Nishizawa et al. 2008).

In addition, to improve levels of thermotolerance, "omics" technology has also been widely used, and to provide valuable information regarding breeding programs. Micromics will facilitate the researcher for the better perceptive of tolerance, and endogenous small noncoding RNAs/miRNA play a major functional role in such a study (Hasanuzzaman et al. 2013; Hemantaranjan et al. 2014). It has been reported that miRNAs regulate various abiotic stress responses by regulating TFs and genes. A number of HS-responsive miRNAs have already been identified in plants, and their roles involved in HS tolerance have been recognized (Yu et al. 2012). During HS, these miRNAs work as negative regulators of stress tolerance as the stress-imposed up-regulated miRNAs may down-regulate their target genes, and also act as positive regulators by the accrual of their target mRNAs gene via stress-prompted, down-regulated miRNAs. For instance, most commonly studied miRNA is miR398, which is involved in responses to various abiotic stresses, mostly HS. This miR398 and its target *CSDs* (closely related copper/zinc superoxide dismutase) were reported in *Brassica rapa* in response to HS (Yu et al. 2012), signifying that in plants the miR398-*CSD*/*CCS* pathways are extensively involved in HS responses. In *H. vulgare*, up-regulation of miR166a was noticed; however, its targets, namely *PHV* (*PHAVOLUTA*), *HOX9* (homeobox leucine zipper protein *HOX9*-like), and *REV* (*REVOLUTA*) were down-regulated in response to elevated temperatures (Kruszka et al. 2014). Chen et al. (2012) document down-regulation of miR168 in *Populus tomentosa* in response to HS that resulted into accrual of its target gene *ARGONAUTE1* (*AGO1*). Hence, overexpression of miRNA-resistant target genes can be helpful to prevail post-transcriptional gene silencing and can result in better expression of engineered traits in transgenic plants (Hasanuzzaman et al. 2013). Therefore, carrying out more research into the role of miRNAs in HS tolerance in plants will help plant scientists genetically engineer stress-tolerant varieties of various crops.

5.7 Conclusions and Future Perspectives

Globally, HS greatly interferes with crop production as it severely influences the growth, development, and productivity of plants. Hence, it is essential to understand the responses of plants and their acclimatization processes to extreme temperatures, and the mechanisms underlying the improvement in the generation of HS-tolerant varieties, especially for agronomically important crops. Currently, several genes have been identified in various studies; however, the information regarding the transcriptional control of high-temperature responses is scarce. Responses of plants to HS show a discrepancy within the species, and also at different developmental stages. Molecular strategies/approaches that reveal the responses and modes of tolerance will be supportive to produce HS-tolerant plants and could be the core for developing crop varieties that can produce economic yields under HS. Also, in the field, modulating cultural practices like selection of cultivars and species, timing and way of seed sowing, and an effective management of irrigation program can significantly diminish the adverse affects of HS.

Recently, external applications of a few of the protectants, like osmoregulators, phytohormones, signaling molecules, etc., have exhibited their profitable effects on plants exposed to extreme temperature, owing to the growth promoting and antioxidant activities of these components. Sugars such as glucose, sucrose, fructan, raffinose, and trehalose not only act as structural constituents of the cells and metabolic resources but also functions as signaling molecules regulating different genes associated with growth and development of plants, and as protectants in response to abiotic stresses like HS. To understand the fundamental mechanisms regarding protective roles of sugars, several efforts have been made against high temperature and also to produce crop varieties with improved HS-tolerance by altering their biosynthetic pathway. Engineering plants to produce these compounds may be a substitute of developing thermotolerant varieties of crops and signifies probably an important part of research on HS tolerance. However, most studies looking to alleviate the impact of HS are restricted to the laboratory. Therefore, agricultural or plant scientists must look to translate the accessible "omics" information into developing HS-tolerant crop varieties at the genomic, proteomic, and/or metabolomic level, and also explore the actual HS responses and their effects on final crop productivity and food safety by using different molecular approaches and agronomic practices at the field level.

References

Akman, Z. (2009). Comparison of high temperature tolerance in maize, rice and sorghum seeds by plant growth regulators. *Journal of Animal and Veterinary Advances* 8: 358–361.

Allakhverdiev, S.I., Kreslavski, V.D., Klimov, V.V. et al. (2008). Heat stress: an overview of molecular responses in photosynthesis. *Photosynthesis Research* 98: 541–550.

Anon, S., Fernandez, J.A., Franco, J.A. et al. (2004). Effects of water stress and night temperature preconditioning on water relations and morphological and anatomical changes of *Lotus creticus* plants. *Scientia Horticulturae* 101: 333–342.

Ashraf, M. and Hafeez, M. (2004). Thermotolerance of pearl millet and maize at early growth stages: growth and nutrient relations. *Biologia Plantarum* 48: 81–86.

Bakhsh, A. and Hussain, T. (2015). Engineering crop plants against abiotic stress: current achievements and prospects. *Emirates Journal of Food and Agriculture* 27: 24–39.

Banowetz, G.M., Ammar, K., and Chen, D.D. (1999). Temperature effects on cytokinin accumulation and kernel mass in dwarf wheat. *Annals of Botany* 83: 303–307.

Barnabás, B., Jäger, K., and Fehér, A. (2008). The effect of drought and heat stress on reproductive processes in cereals. *Plant, Cell & Environment* 31: 11–38.

Bieleski, R.L. (1982). Sugar alcohols. In: *Encyclopedia of Plant Physiology*, Plant Carbohydrates I, New Series, vol. 13A (eds. F.A. Loewus and W. Tanner), 158–192. Berlin: Springer-Verlag.

Biłas, R., Szafran, K., Hnatuszko-Konka, K., and Kononowicz, A.K. (2016). *Cis*-regulatory elements used to control gene expression in plants. *Plant Cell Tissue and Organ Culture* 127: 269–287.

Bita, C.E. and Gerats, T. (2013). Plant tolerance to high temperature in a changing environment: scientific fundamentals and production of heat stress-tolerant crops. *Frontiers in Plant Science* 4: 1–18.

Busch, W., Wunderlich, M., and Schoffl, F. (2005). Identification of novel heat shock factor-dependent genes and biochemical pathways in *Arabidopsis thaliana*. *The Plant Journal* 41: 1–14.

Camejo, D., Jiménez, A., Alarcón, J.J. et al. (2006). Changes in photosynthetic parameters and antioxidant activities following heat-shock treatment in tomato plants. *Functional Plant Biology* 33: 177–187.

Cao, Y.Y., Duan, H., Yang, L.N. et al. (2008). Effect of heat stress during meiosis on grain yield of rice cultivars differing in heat tolerance and its physiological mechanism. *Acta Agronomica Sinica* 34: 2134–2142.

Chandrakar, V. and Keshavkant, S. (2018). Nitric oxide and dimethylthiourea up-regulates pyrroline-5-carboxylate synthetase expression to improve arsenic tolerance in *Glycine max* L. *Environmental Progress & Sustainable Energy* 38 (2): 402–409.

Chandrakar, V., Naithani, S.C., and Keshavkant, S. (2016). Arsenic-induced metabolic disturbances and their mitigation mechanisms in crop plants: a review. *Biologia* 71: 367–377.

Chandrakar, V., Dubey, A., and Keshavkant, S. (2018). Modulation of arsenic-induced oxidative stress and protein metabolism by diphenyleneiodonium, 24-epibrassinolide and proline in *Glycine max* L. *Acta Botanica Croatica* 77: 51–61.

Chen, L., Ren, Y., Zhang, Y. et al. (2012). Genome-wide identification and expression analysis of heat-responsive and novel microRNAs in *Populus tomentosa*. *Gene* 504: 160–165.

Cheng, L., Zou, Y., Ding, S. et al. (2009). Polyamine accumulation in transgenic tomato enhances the tolerance to high temperature stress. *Journal of Integrative Plant Biology* 51: 489–499.

Crafts-Brandner, S.J. and Salvucci, M.E. (2002). Sensitivity of photosynthesis in a C4 plant, maize, to heat stress. *Plant Physiology* 129: 1773–1780.

Delorge, I., Janiak, M., Carpentier, S., and Dijck, P.V. (2014). Fine tuning of trehalose biosynthesis and hydrolysis as novel tools for the generation of abiotic stress tolerant plants. *Frontiers in Plant Science* 5: 1–9.

Delvalle, D., Dumez, S., Wattebled, F. et al. (2005). Soluble starch synthase I: a major determinant for the synthesis of amylopectin in *Arabidopsis thaliana* leaves. *The Plant Journal* 43: 398–412.

Djanaguiraman, M., Sheeba, J.A., Devi, D.D., and Bangarusamy, U. (2009). Cotton leaf senescence can be delayed by nitrophenolate spray through enhanced antioxidant defense system. *Journal of Agronomy and Crop Science* 195: 213–224.

Djanaguiraman, M., Prasad, P.V.V., and Seppanen, M. (2010). Selenium protects sorghum leaves from oxidative damage under high temperature stress by enhancing antioxidant defense system. *Plant Physiology and Biochemistry* 48: 999–1007.

Djanaguiraman, M., Prasad, P.V.V., Boyle, D.L., and Schapaugh, W.T. (2011). High-temperature stress and soybean leaves: leaf anatomy and photosynthesis. *Crop Science* 51: 2125–2131.

Doehlemann, G., Berndt, P., and Hahn, M. (2006). Trehalose metabolism is important for heat stress tolerance and spore germination of *Botrytis cinerea*. *Microbiology* 152: 2625–2634.

Dubey, R.S. (2005). Photosynthesis in plants under stressful conditions. In: *Handbook of Photosynthesis* (ed. M. Pessarakli), 717–737. Boca Raton, FL: CRC Press.

Ebrahim, M.K., Zingsheim, O., El-Shourbagy, M.N. et al. (1998). Growth and sugar storage in sugarcane grown at temperature below and above optimum. *Journal of Plant Physiology* 153: 593–602.

Essemine, J., Ammar, S., and Bouzid, S. (2010). Impact of heat stress on germination and growth in higher plants: physiological, biochemical and molecular repercussions and mechanisms of defence. *Journal of Biological Sciences* 10: 565–572.

Foolad, M.R. (2005). Breeding for abiotic stress tolerances in tomato. In: *Abiotic Stresses: Plant Resistance through Breeding and Molecular Approaches* (eds. M. Ashraf and P.J.C. Harris), 613–684. New York: The Haworth Press Inc.

Galtier, N., Foyer, C.H., Huber, J. et al. (1993). Effect of elevated sucrose phosphate synthase activity on photo-synthesis, assimilate partitioning, and growth in tomato (*Lycopersicon esculentum* var UC82B). *Plant Physiology* 101: 535–543.

Geigenberger, P., Merlo, L., Reimholz, R., and Stitt, M. (1994). When growing potato tubers are detached from their mother plant there is a rapid inhibition of starch synthesis, involving inhibition of ADP-glucose pyrophosphorylase. *Planta* 193: 486–493.

Granot, D., David-Schwartz, R., and Kelly, G. (2013). Hexose kinases and their role in sugar sensing and plant development. *Frontiers in Plant Science* 4: 44.

Greer, D.H. and Weedon, M.M. (2012). Modelling photosynthetic responses to temperature of grapevine (*Vitis vinifera* cv. Semillon) leaves on vines grown in a hot climate. *Plant, Cell & Environment* 35: 1050–1064.

Gunawardhana, M.D.M. and de Silva, C.S. (2011). Impact of temperature and water stress on growth yield and related biochemical parameters of okra. *Tropical Agricultural Research* 23: 77–83.

Guo, M., Liu, J.-H., Ma, X. et al. (2016). The plant heat stress transcription factors (HSFs): structure, regulation, and function in response to abiotic stresses. *Frontiers in Plant Science* 7: 114.

Gupta, A.K. and Kaur, N. (2005). Sugar signalling and gene expression in relation to carbohydrate metabolism under abiotic stresses in plants. *Journal of Biosciences* 30: 761–776.

Gupta, K., Agarwal, P.K., Reddy, M.K., and Jha, B. (2010). SbDREB2A, an A-2 type DREB transcription factor from extreme halophyte *Salicornia brachiata* confers abiotic stress tolerance in *Escherichia coli*. *Plant Cell Reports* 29: 1131–1137.

Harsh, A., Sharma, Y.K., Joshi, U. et al. (2016). Effect of short-term heat stress on total sugars, proline and some antioxidant enzymes in moth bean (*Vigna aconitifolia*). *Annals of Agricultural Science* 61: 57–64.

Hasanuzzaman, M., Hossain, M.A., and Fujita, M. (2010). Physiological and biochemical mechanisms of nitric oxide induced abiotic stress tolerance in plants. *American Journal of Plant Science* 5: 295–324.

Hasanuzzaman, M., Nahar, K., Alam, M.M. et al. (2013). Physiological, biochemical, and molecular mechanisms of heat stress tolerance in plants. *International Journal of Molecular Sciences* 14: 9643–9684.

Hedhly, A. (2011). Sensitivity of flowering plant gametophyte to temperature fluctuations. *Environmental and Experimental Botany* 74: 9–16.

Hemantaranjan, A., Nishant Bhanu, A., Singh, M.N. et al. (2014). Heat stress responses and thermotolerance. *Advances in Plants & Agriculture Research* 1: 62–70.

Hogy, P., Poll, C., Marhan, S. et al. (2013). Impacts of temperature increase and change in precipitation pattern on crop yield and yield quality of barley. *Food Chemistry* 136: 1470–1477.

Hong, S.W. and Vierling, E. (2000). Mutants of *Arabidopsis thaliana* defective in the acquisition of tolerance to high temperature stress. *Proceedings of the National Academy of Sciences of the United States of America* 97: 4392–4397.

Hua, J. (2009). From freezing to scorching transcriptional responses to temperature variations in plants. *Current Opinion in Plant Biology* 12: 568–573.

Huang, Y.W., Nie, Y.X., Wan, Y.Y. et al. (2013). Exogenous glucose regulates activities of antioxidant enzyme, soluble acid invertase and neutral invertase and alleviates dehydration stress of cucumber seedlings. *Scientia Horticulturae* 162: 20–30.

Huang, Y.W., Zhou, Z.Q., Yang, H.X. et al. (2015). Glucose application protects chloroplast ultrastructure in heat-stressed cucumber leaves through modifying antioxidant enzyme activity. *Biologia Plantarum* 59: 131–138.

Hurkman, W.J., Vensel, W.H., Tanaka, C.K. et al. (2009). Effect of high temperature on albumin and globulin accumulation in the endosperm proteome of the developing wheat grain. *Journal of Cereal Science* 49: 12–23.

Iba, K. (2002). Acclimative response to temperature stress in higher plants: approaches of gene engineering for temperature tolerance. *Annual Review of Plant Biology* 53: 225–245.

Janska, A., Marsik, P., Zelenkova, S., and Ovesna, J. (2010). Cold stress and acclimation: what is important for metabolic adjustment? *Plant Biology* 12: 395–405.

Jha, U., Bohra, A., and Singh, N.P. (2014). Heat stress in crop plants: its nature, impacts and integrated breeding strategies to improve heat tolerance. *Plant Breeding* 133: 679–701.

Kalra, N., Chakraborty, D., Sharma, A. et al. (2008). Effect of increasing temperature on yield of some winter crops in northwest India. *Current Science* 94: 82–88.

Kaplan, F., Kopka, J., Haskell, D.W. et al. (2004). Exploring the temperature-stress metabolome of *Arabidopsis*. *Plant Physiology* 136: 4159–4168.

Karim, M.A., Fracheboud, Y., and Stamp, P. (1997). Heat tolerance of maize with reference of some physiological characteristics. *Annals of Bangladesh Agriculture* 7: 27–33.

Kaur, P., Ghai, N., and Sangha, M.K. (2009). Induction of thermo tolerance through heat acclimation and salicylic acid in *Brassica species*. *African Journal of Biotechnology* 8: 619–625.

Kempa, S., Krasensky, J., Dal Santo, S. et al. (2008). A central role of abscisic acid in stress-regulated carbohydrate metabolism. *PLoS One* 3: 3935.

Kepova, K.D., Holzer, R., Stoilova, L.S., and Feller, U. (2005). Heat stress effects on ribulose-1,5-bisphosphate carboxylase/oxygenase, Rubisco binding protein and Rubisco activase in wheat leaves. *Biologia Plantarum* 49: 521–525.

Khan, M.S., Ahmad, D., and Khan, M.A. (2015). Utilization of genes encoding osmoprotectants in transgenic plants for enhanced abiotic stress tolerance. *Electronic Journal of Biotechnology* 18: 257–266.

Krasensky, J. and Jonak, C. (2012). Drought, salt, and temperature stress-induced metabolic rearrangements and regulatory networks. *Journal of Experimental Botany* 63: 1593–1608.

Kruszka, K., Pacak, A., Swidabarteczka, A. et al. (2014). Transcriptionally and post-transcriptionally regulated microRNAs in heat stress response in barley. *Journal of Experimental Botany* 65: 6123–6135.

Kumar, S., Sairam, R.K., and Prabhu, K.V. (2013). Physiological traits for high temperature stress tolerance in *Brassica juncea*. *Indian Journal of Plant Physiology* 18: 89–93.

Lafta, A.M. and Lorenzen, J.H. (1995). Effect of high temperature on plant growth and carbohydrate metabolism in potato. *Plant Physiology* 109: 637–643.

Lamaoui, M., Martin Jemo, M., Datla, R., and Bekkaoui, F. (2018). Heat and drought stresses in crops and approaches for their mitigation. *Frontiers in Chemistry* 6: 26.

Lee, B.H., Won, S.H., Lee, H.S. et al. (2000). Expression of the chloroplast-localized small heat shock protein by oxidative stress in rice. *Gene* 245: 283–290.

Lee, J.H., Hubel, A., and Schoffl, F. (1995). Derepression of the activity of genetically engineered heat shock factor causes constitutive synthesis of heat shock proteins and increased thermotolerance in transgenic *Arabidopsis*. *The Plant Journal* 8: 603–612.

Li, L. and Sheen, J. (2016). Dynamic and diverse sugar signaling. *Current Opinion in Plant Biology* 33: 116–125.

Li, X., Lawas, L.M.F., Malo, R. et al. (2015). Metabolic and transcriptomic signatures of rice floral organs reveal sugar starvation as a factor in reproductive failure under heat and drought stress. *Plant, Cell & Environment* 38: 2171–2192.

Liu, C., Zhao, L., and Yu, G. (2011). The dominant glutamic acid metabolic flux to produce gamma-amino butyric acid over proline in *Nicotiana tabacum* leaves under water stress relates to its significant role in antioxidant activity. *Journal of Integrative Plant Biology* 53: 608–618.

Liu, X. and Huang, B. (2000). Carbohydrate accumulation in relation to heat stress tolerance in two creeping bentgrass cultivar. *Journal of the American Society for Horticultural Science* 125: 442–447.

Livingston, D.P., Hincha, D.K., and Heyer, A.G. (2009). Fructan and its relationship to abiotic stress tolerance in plants. *Cellular and Molecular Life Sciences* 66: 2007–2023.

Lobell, D.B., Hammer, G.L., McLean, G. et al. (2013). The critical role of extreme heat for maize production in the United States. *Nature Climate Change* 3: 497–501.

Los, D.A. and Murata, N. (2004). Membrane fluidity and its roles in the perception of environmental signals. *Biochimica et Biophysica Acta* 1666: 142–157.

Luo, Y., Gao, Y.M., Wang, W., and Zou, C.J. (2014). Application of trehalose ameliorates heat stress and promotes recovery of winter wheat seedlings. *Biologia Plantarum* 58: 395–407.

Ma, H., Wang, C., Yang, B. et al. (2016). CarHSFB2, a class B heat shock transcription factor, is involved in different developmental processes and various stress responses in chickpea (*Cicer arietinum* L.). *Plant Molecular Biology Reporter* 34: 1–14.

Machado, S. and Paulsen, G.M. (2001). Combined effects of drought and high temperature on water relations of wheat and sorghum. *Plant and Soil* 233: 179–187.

Maheswari, M., Yadav, S.K., Shanker, A.K. et al. (2012). Overview of plant stresses: mechanisms, adaptations and research pursuit. In: *Crop Stress and its Management: Perspectives and Strategies* (eds. B. Venkateswarlu, A.K. Shanker, C. Shanker and M. Maheswari), 1–8. Dordrecht, The Netherlands: Springer.

Marias, D., Meinzer, F.C., and Still, C. (2017). Impacts of leaf age and heat stress duration on photosynthetic gas exchange and foliar nonstructural carbohydrates in *Coffea arabica*. *Ecology and Evolution* 7: 1297–1310.

Mazorra, L.M., Nunez, M., Echerarria, E. et al. (2002). Influence of brassinosteroids and antioxidant enzymes activity in tomato under different temperatures. *Plant Biology* 45: 593–596.

Miranda, J.A., Avonce, N., Suárez, R. et al. (2007). A bifunctional TPS-TPP enzyme from yeast confers tolerance to multiple and extreme abiotic-stress conditions in transgenic *Arabidopsis*. *Planta* 226: 1411–1421.

Mohammed, A.R. and Tarpley, L. (2010). Effects of high night temperature and spikelet position on yield-related parameters of rice (*Oryza sativa* L.) plants. *European Journal of Agronomy* 33: 117–123.

Morales, D., Rodriguez, P., Dellamico, J. et al. (2003). High-temperature preconditioning and thermal shock imposition affects water relations, gas exchange and root hydraulic conductivity in tomato. *Biologia Plantarum* 47: 203–208.

Nadeem, M., Li, J., Wang, M. et al. (2018). Unraveling field crops sensitivity to heat stress: mechanisms, approaches, and future prospects. *Agronomy* 8: 1–34.

Nishizawa, A., Yabuta, Y., and Shigeoka, S. (2008). Galactinol and raffinose constitute a novel function to protect plants from oxidative damage. *Plant Physiology* 147: 1251–1263.

Obaid, A.Y., Sabir, J.S.M., Atef, A. et al. (2016). Analysis of transcriptional response to heat stress in *Rhazya stricta*. *BMC Plant Biology* 16: 252.

Parrotta, L., Faleri, C., Cresti, M., and Cai, G. (2016). Heat stress affects the cytoskeleton and the delivery of sucrose synthase in tobacco pollen tubes. *Planta* 243: 43–63.

Paul, M.J., Primavesi, L.F., Jhurreea, D., and Zhang, Y.H. (2008). Trehalose metabolism and signaling. *Annual Review of Plant Biology* 59: 417–441.

Peterbauer, T. and Richter, A. (2001). Biochemistry and physiology of raffinose family oligosaccharides and galactosyl cyclitols in seeds. *Seed Science Research* 11: 185–197.

Piramila, B.H.M., Prabha, A.L., Nandagopalan, V., and Stanley, A.L. (2012). Effect of heat treatment on germination, seedling growth and some biochemical parameters of dry seeds of black gram. *International Journal of Pharmaceutical and Phytopharmacological Research* 1: 194–202.

Potters, G., Pasternak, T.P., Guisez, Y., and Jansen, M.A.K. (2008). Different stresses similar morphogenic responses integrating a plethora of pathways. *Plant, Cell & Environment* 32: 158–169.

Preiss, J. (1982). Regulation of the biosynthesis and degradation of starch. *Annual Review of Plant Physiology* 33: 431–454.

Ren, C., Bilyeu, K.D., and Beuselinck, P. (2009). Composition, vigor, and proteome of mature soybean seeds developed under high temperature. *Crop Science* 49: 1010–1022.

Ruan, Y.L. (2014). Sucrose metabolism: gateway to diverse carbon use and sugar signalling. *Annual Review of Plant Biology* 65: 33–67.

Ruan, Y.L., Jin, Y., Yang, Y.J. et al. (2010). Sugar input metabolism and signaling mediated by invertase roles in development yield potential and response to drought and heat. *Molecular Plant* 3: 942–955.

Saidi, Y., Finka, A., Muriset, M. et al. (2009). The heat shock response in moss plants is regulated by specific calcium-permeable channels in the plasma membrane. *Plant Cell* 21: 2829–2843.

Saini, H.S. (1997). Effects of water stress on male gametophyte development in plants. *Sexual Plant Reproduction* 10: 67–73.

Sakata, T. and Higashitani, A. (2008). Male sterility accompanied with abnormal anther development in plants: genes and environmental stresses with special reference to high temperature injury. *International Journal of Plant Developmental Biology* 2: 42–51.

Salvucci, M.E. and Crafts-Brandner, S.J. (2004). Inhibition of photosynthesis by heat stress: the activation state of Rubisco as a limiting factor in photosynthesis. *Physiologia Plantarum* 120: 179–186.

Sanmiya, K., Suzuki, K., Egawa, Y., and Shono, M. (2004). Mitochondrial small heat shock protein enhances thermotolerance in tobacco plants. *FEBS Letters* 557: 265–268.

Sato, S., Kamiyama, M., Iwata, T. et al. (2006). Moderate increase of mean daily temperature adversely affects fruit set of *Lycopersicon esculentum* by disrupting specific physiological processes in male reproductive development. *Annals of Botany* 97: 731–738.

Savchenko, G.E., Klyuchareva, E.A., Abrabchik, L.M., and Serdyuchenko, E.V. (2002). Effect of periodic heat shock on the membrane system of etioplasts. *Russian Journal of Plant Physiology* 49: 349–359.

Sayed, O.H. (1996). Adaptational responses of *Zygophyllum qatarense* Hadidi to stress conditions in a desert environment. *Journal of Arid Environments* 32: 445–452.

Schrader, S.M., Wise, R.R., Wacholtz, W.F. et al. (2004). Thylakoid membrane responses to moderately high leaf temperature in Pima cotton. *Plant, Cell & Environment* 27: 725–735.

Shah, N. and Paulsen, G. (2003). Interaction of drought and high temperature on photosynthesis and grain-filling of wheat. *Plant and Soil* 257: 219–226.

Sheoran, I.S. and Saini, H.S. (1996). Drought-induced male sterility in rice: changes in carbohydrate levels and enzyme activities associated with the inhibition of starch accumulation in pollen. *Sexual Plant Reproduction* 9: 161–169.

Shinozaki, K., Yamaguchi-Shinozaki, K., and Seki, M. (2003). Regulatory network of gene expression in the drought and cold stress responses. *Current Opinion in Plant Biology* 6: 410–417.

Simoes-Araujo, J.L., Rumjanek, N.G., and Margis-Pinheiro, M. (2003). Small heat shock proteins genes are differentially expressed in distinct varieties of common bean. *Brazilian Journal of Plant Physiology* 15: 33–41.

Sinsawat, V., Leipner, J., Stamp, P., and Fracheboud, Y. (2004). Effect of heat stress on the photosynthetic apparatus in maize (*Zea mays* L.) grown at control or high temperature. *Environmental and Experimental Botany* 52: 123–129.

Smirnoff, N. and Cumbes, Q.J. (1989). Hydroxyl radical scavenging activity of compatible solutes. *Phytochemistry* 28: 1057–1060.

Stark, D.M., Timmerman, K.P., Barry, G.F. et al. (1992). Regulation of the amount of starch in plant tissues by ADP glucose pyrophosphorylase. *Science* 258: 287–292.

Suárez, R., Calderón, C., and Iturriaga, G. (2009). Improved tolerance to multiple abiotic stresses in transgenic alfalfa accumulating trehalose. *Crop Science* 49: 1791–1799.

Sumesh, K.V., Sharma-Natu, P., and Ghildiyal, M.C. (2008). Starch synthase activity and heat shock protein in relation to thermal tolerance of developing wheat grains. *Biologia Plantarum* 52: 749–753.

Suzuki, K., Takeda, H., Tsukaguchi, T., and Egawa, Y. (2001). Ultrastructural study on degeneration of tapetum in anther of snapbean (*Phaseolus vulgaris* L.) under heat stress. *Sexual Plant Reproduction* 13: 293–299.

Suzuki, N., Rizhsky, L., Liang, H. et al. (2005). Enhanced tolerance to environmental stress in transgenic plants expressing the transcriptional coactivator multiprotein bridging factor 1c[1[w]]. *Plant Physiology* 139: 1313–1322.

Suzuki, N., Bajad, S., Shuman, J. et al. (2008). The transcriptional co-activator MBF1c is a key regulator of thermotolerance in *Arabidopsis thaliana*. *Journal of Biological Chemistry* 283: 9269–9275.

Suzuki, N., Koussevitzky, S., Mittler, R., and Miller, G. (2012). ROS and redox signalling in the response of plants to abiotic stress. *Plant, Cell & Environment* 35: 259–270.

Tayade, R., Nguyen, T.D., Oh, S.A. et al. (2018). Effective strategies for enhancing tolerance to high-temperature stress in rice during the reproductive and ripening stages. *Plant Breed Biotech* 6: 1–18.

Tian, J., Belanger, F.C., and Huang, B. (2009). Identification of heat stress responsive genes in heat-adapted thermal *Agrostis scabra* by suppression subtractive hybridization. *Journal of Plant Physiology* 166: 588–601.

Todorov, D.T., Karanov, E.N., Smith, A.R., and Hall, M.A. (2003). Chlorophyllase activity and chlorophyll content in wild type and *eti 5* mutant of *Arabidopsis thaliana* subjected to low and high temperatures. *Biologia Plantarum* 46: 633–636.

Tsukaguchi, T., Kawamitsu, Y., Takeda, H. et al. (2003). Water status of flower buds and leaves as affected by high temperature in heat tolerant and heat-sensitive cultivars of snap bean (*Phaseolus vulgaris* L.). *Plant Production Science* 6: 4–27.

Valluru, R. and Van den Ende, W. (2008). Plant fructans in stress environments: emerging concepts and future prospects. *Journal of Experimental Botany* 59: 2905–2916.

Wahid, A. and Close, T.J. (2007). Expression of dehydrins under heat stress and their relationship with water relations of sugarcane leaves. *Biologia Plantarum* 51: 104–109.

Wahid, A. and Shabbir, A. (2005). Induction of heat stress tolerance in barley seedlings by presowing seed treatment with glycine betaine. *Plant Growth Regulation* 46: 133–141.

Wahid, A., Gelani, S., Ashraf, M., and Foolad, M.R. (2007). Heat tolerance in plants: an overview. *Environmental and Experimental Botany* 61: 199–223.

Wallwork, M.A.B., Logue, S.J., MacLeod, L.C., and Jenner, C.F. (1998). Effect of high temperature during grain filling on starch synthesis in the developing barley grain. *Australian Journal of Plant Physiology* 25: 173–181.

Wang, J.Z., Cui, L.J., Wang, Y., and Li, J.L. (2009). Growth, lipid peroxidation and photosynthesis in two tall fescue cultivars differing in heat tolerance. *Biologia Plantarum* 53: 247–242.

Wang, Q.L., Chen, J.H., He, N.Y., and Guo, F.Q. (2018). Metabolic reprogramming in chloroplasts under heat stress in plants. *International Journal of Molecular Sciences* 19: 849.

Warland, J.S., McDonald, M.R., and McKeown, A.M. (2006). Annual yields of five crops in the family Brassicaceae in southern Ontario in relation to weather and climate. *Canadian Journal of Plant Science* 86: 1209–1215.

Wolf, S., Marani, A., and Rudich, J. (1990). Effects of temperature and photoperiod on assimilate partitioning in potato plants. *Annals of Botany* 66: 513–520.

Wu, X., Shiroto, Y., Kishitani, S. et al. (2009). Enhanced heat and drought tolerance in transgenic rice seedlings overexpressing *OsWRKY11* under the control of *HSP101* promoter. *Plant Cell Reports* 28: 21–30.

Xu, D.P., Sun, S.S., and Black, C.C. (1989). Sucrose metabolism in lima bean seeds. *Plant Physiology* 89: 1106–1116.

Xu, J., Xue, C., Xue, D. et al. (2013). Overexpression of *GmHsp90s*, a heat shock protein 90 (Hsp90) gene family cloning from soybean, decrease damage of abiotic stresses in *Arabidopsis thaliana*. *PLoS One* 8: e69810.

Yadu, B., Chandrakar, V., and Keshavkant, S. (2016). Responses of plants towards fluoride: an overview of oxidative stress and defense mechanisms. *Fluoride* 49: 293–302.

Yang, J.D., Yun, J.Y., Zhang, T.H., and Zhao, H.L. (2006). Presoaking with nitric oxide donor SNP alleviates heat shock damages in mung bean leaf discs. *Botanical Studies* 47: 129–136.

Yu, X., Wang, H., Lu, Y. et al. (2012). Identification of conserved and novel microRNAs that are responsive to heat stress in *Brassica rapa*. *Journal of Experimental Botany* 63: 1025–1038.

Yuan, L., Tang, L., Zhu, S. et al. (2017). Influence of heat stress on leaf morphology and nitrogen–carbohydrate metabolisms in two wucai (*Brassica campestris* L.) genotypes. *Acta Societatis Botanicorum Poloniae* 86: 3554.

Zhang, J., Huang, W., Pan, Q., and Liu, Y. (2005). Improvement of chilling tolerance and accumulation of heat shock protein sin grape berries (*Vitis vinifera* cv. Jingxiu) by heat pretreatment. *Postharvest Biology and Technology* 38: 80–90.

Zhang, X., Cai, J., Wollenweber, B. et al. (2013). Multiple heat and drought events affect grain yield and accumulations of high molecular weight glutenin subunits and glutenin macropolymers in wheat. *Journal of Cereal Science* 57: 134–140.

Zhang, Y., Mian, M.A.R., and Bouton, J.H. (2006). Recent molecular and genomic studies on stress tolerance of forage and turf grasses. *Crop Science* 46: 497–511.

Zhou, Z., Yuan, Y., Zhou, W., and Zhang, C. (2016). Effects of exogenously supplied sucrose on OsSUTs and OsSPSs transcript abundances and rice root ammonium assimilation. *Acta Physiologiae Plantarum* 38: 274.

Zuther, E., Buchel, K., Hundertmark, M. et al. (2004). The role of raffinose in the cold acclimation response of *Arabidopsis thaliana*. *FEBS Letters* 576: 167–173.

6

Transcriptomics to Dissect Plant Responses to Heat Stress

Sagar Satish Datir

[1] *Biology Department, Queen's University, Kingston, ON, Canada*
[2] *Department of Biotechnology, Savitribai Phule Pune University, Pune, MS, India*

6.1 Introduction

Climate change resulting in temperature fluctuations is one of the major abiotic stresses affecting the growth and productivity of agriculturally important crops. It is evident that future crop production will mainly be threatened and limited by global warming which predicts an increase of 2–6 °C in ambient temperature by the end of the twenty-first century (Siebert and Ewert 2014; Yan et al. 2016). Climate change is inevitable, and hence Christensen and Christensen (2007) suggest that it is already having a very significant influence on agricultural production worldwide as heatwaves have resulted in severe crop yield losses causing great risk for future global food security. Owing to their sessile nature, plants have adapted various mechanisms to tolerate heat stress, such as avoidance and tolerance (Sailaja et al. 2014). Plant adaptations to a wide range of temperatures include reprogramming their transcriptome, proteome, and metabolome as well as activating cell death mechanisms, which further leads to organ abortion or entire plant death (Qi et al. 2011; Sánchez-Rodríguez et al. 2011). Both vegetative as well as reproductive organs have been found to be affected by heat stress. However, it severely affects the development of reproductive organs such as inflorescences and floral

Heat Stress Tolerance in Plants: Physiological, Molecular and Genetic Perspectives, First Edition.
Edited by Shabir Hussain Wani and Vinay Kumar.

organs (Suzuki et al. 2014; Thakur et al. 2010; Wu et al. 2009). Plant responses to heat stress greatly vary among crop species and disturb various cellular, developmental, biochemical, and molecular processes. Although various physiological, biochemical, and molecular mechanisms underlying stress tolerance in plant are well studied (Hasanuzzaman et al. 2013a,2013b), at the molecular level it is necessary to understand the expression and function of various genes which up-regulate or down-regulate in response to high temperature. As heat stress is becoming a severe issue for agriculture, it is important to identify various molecular mechanisms that drive the response of plants to heat stress (Chen and Li 2017) and develop heat-tolerant plant cultivars. Conventional breeding approaches for the production of heat-tolerant lines are underway; however, the genetic and physiological basis behind improvements remains unclear. Moreover, these approaches can be slow mainly because of a variable climate (Cottee et al. 2014). Therefore, Tardieu and Hammer (2012) suggest that to provide the foundation for a trait-based approach for breeding of plants with superior physiological and yield performance under heat stress conditions a complete understanding and knowledge of molecular mechanisms are essential. Molecular analysis of plant responses to heat stress revealed that it is a very complex and varied phenomenon which involves complex multigene interactions (Driedonks et al. 2015). Omics technologies provide new avenues for the identification of transcriptional, translational, and post-translational mechanisms and signaling pathways in response to various abiotic stresses, including heat stress (Aprile et al. 2009; Hasanuzzaman et al. 2013b). Moreover, omics-based platforms have also increased the availability of genomic and transcriptomic resources that are very crucial resources which not only promote research in model and applied plant species but also allow researchers to discover and isolate important genes and to analyze their functions that regulate yields and tolerance to various environmental stresses (Mochida and Shinozaki 2010). Transcriptomics allow studying an organism's transcriptome, i.e. a complete set of expressed genes especially through mRNA profiling in different tissues, under a given time and a particular condition. It provides a snapshot of the total transcripts present in a cell, which gives valuable information on how various genes are regulated and also to explain their functions (García-Ortega and Martínez 2015; Lowe et al. 2017). Transcriptomics provides detailed information on various transcription factors (TFs) involved in that specific condition. For instance, several studies reveal the induction of heat shock proteins (HSPs), heat shock factors (HSFs), up-regulation of genes involved in synthesis of antioxidants, and osmolytes under heat stress (Fang et al. 2018; Galsurker et al. 2018; Qin et al. 2008; Y. Wang et al. 2016). As a result, plants produce increased levels of antioxidant compounds, and there is more production of osmoprotectants such as proline and glycine betaine and activation of signal transduction cascades through the induction of signaling molecules as adaptation strategies. However, plants do differ in adaptation strategies and stress tolerance capacity under heat stress, which is mainly due to the differences in their genetic constitution that further leads to the change in associated heat-stress tolerance gene expression as well as small RNA (sRNA) involved in synthesis of metabolites. Hence, it is very crucial to understand heat response at the molecular level via the transcriptome route to unravel the complex mechanisms behind heat-stress tolerance in various crop plants (Sailaja et al. 2014). There have been very few reports on transcriptome studies conducted on major crop plants in response to heat stress. This chapter briefly summarizes the information on transcriptome studies conducted in some crop plants. Table 6.1 describes key genes identified in transcriptome studies conducted in various plants in response to heat stress.

6.1.1 Transcriptome Sequencing and Expression Profiling of Genes Involved in Heat Stress Response

6.1.1.1 Rice (*Oryza sativa* L.)

Rice (*Oryza sativa* L.) is a major cereal crop worldwide severely affected by heat stress in tropical and subtropical regions (Ishimaru et al. 2015). High temperatures coinciding with reproductive stages/flowering stages such as anthesis/grain filling prompt major yield loss in rice (Bheemanahalli et al. 2016; Jagadish et al. 2010; Sathishraj et al. 2015; Yang et al. 2015). Shah et al. (2011) report that booting and flowering are the

Table 6.1 Key genes identified in transcriptome profiling.

Crop	Tissue	Comment	Reference
Triticum aestivum	Leaves	Up regulation of heat shock proteins, heat shock factors, etc.	Qin et al. (2008)
Triticum aestivum	Leaves	*TaHsp90*	Vishwakarma et al. (2018)
Hordeum vulgare	Caryopses	*HvTPS1* (trehalose-6-phosphate synthase) and *HvTPS2*, etc.	Elke et al. (2011)
Panicum virgatum	Leaves	Up-regulation of ATPase regulator, chaperone binding, and protein folding	Li et al. (2013)
Capsicum annum	Leaves	Upregulation of genes involved in heat shock protein, heat shock transcription factors, hormone, as well as calcium and kinase signaling	Li et al. (2015)
Dianthus caryophyllus	Leaves	Heat shock factors, Heat shock proteins, ROS-scavenging enzymes and genes related to signal transduction and carbohydrate metabolism	Wan et al. (2015)
Oryza sativa	Pistils/ anthers	Up-regulation of BAG6-like, Hsp101, calcyclin-BP and OsSTI1, etc.	González-Schain et al. (2016)
Oryza sativa	Leaves	Heat shock protein family	Wang et al. (2018)
Brassica rapa	Leaves	Induction of heat shock proteins, heat shock factors, etc.	Dong et al. (2015)
Brassica rapa ssp. *chinensis*	Leaves	Up-regulation of transcription factor, genes related to photosynthesis, etc.	A. Wang et al. (2016)

Lolium perenne	Leaves	Heat shock factors	Wang et al. (2016)
Lolium perenne	Leaves	Heat shock factors, heat shock proteins, and antioxidant related genes	Wang et al. (2017)
Spinacia oleracea	Leaves	Up-regulation of heat shock proteins, NBS-LRR type resistance protein, disease resistance protein, etc.	Yan et al. (2013)
Brachypodium distachyon	Leaves and inflores-cences	Mainly transcription factors	Chen and Li (2017)
Zea mays	Leaves	Low levels of genes involved in zeatin and brassinosteroid biosynthesis in heat tolerant genotype	Shi et al. (2017)
Arabidopsis thaliana	Flowers and leaves	Up-regulation of unfold protein response genes	
Onion cepa	Bulbs Outer scale	High expression of genes involved in lipid metabolism, oxidation pathways, and cell-wall modification in the outer scale during heating	Galsurker et al. (2018)
	Bulbs Inner scale	Defense response related genes such as those encoding heat-shock proteins, antioxidative stress defense, or production of osmoprotectant metabolites	
Oryza sativa	Whole plant	High expression of heat shock proteins, beta-carotene hydroxylase, cinnamyl alcohol 3, etc.	Fang et al. (2018)
Musa species AAA Group	Leaves	Heat shock proteins, heat shock factors, glyceraldehyde 3-phosphate, etc.	Vidya et al. (2018)

most critical stages affected by heat stress which may lead to complete sterility in rice. Hence, it is essential to understand the mechanism underlying heat tolerance for genetic improvement in rice. Genome-wide transcriptome analysis of transgenic rice at seedling stage (Jung et al. 2012) and heat tolerant and heat susceptible rice varieties during anthesis stage revealed new insights into the molecular basis of heat stress responses (González-Schain et al. 2016). Transcriptome analysis showed repression of the expression of genes encoding a number of TF families, along with signal transduction and metabolic pathway, while induction of the expression of genes encoding HSFs and HSPs was observed. It is noteworthy that several of these genes were mainly expressed at the late stages of anther development in rice. Induction levels of a set of heat-responsive genes in reproductive tissues were found to be correlated with heat tolerance. These studies also suggested that to overcome stress damage and to ensure fertilization the proper expression of protective chaperones is necessary in anthers before anthesis. As the anthesis stage in rice is the most sensitive to high temperatures (Bheemanahalli et al. 2016; Prasad et al. 2006), the information provided on candidate genes identification in response to heat stress in this study can be considered in molecular breeding programs for the development of highly productive heat-tolerant rice cultivars (González-Schain et al. 2016). Fang et al. (2018) performed heat-stress-responsive transcriptome analysis in heat susceptible and heat tolerant rice breeding lines. Seedlings of both the lines were exposed to heat treatment (at 45 °C for 0, 12 or 24 hours) at three leaf stages. Control (untreated with heat) and treated seedlings were subjected to RNA sequencing to understand the transcriptome differences between heat susceptible and heat tolerant rice breeding lines. Transcriptome analysis of both heat susceptible and heat tolerant lines altogether revealed a total 32 391 unique genes. At 12 hours, a sum of 2275 (heat susceptible) and 2603 (heat tolerant) genes were up-regulated and 2270 genes (heat susceptible) and 2900 (heat tolerant) were down-regulated when compared with 0 hour. On the other hand, as compared to 0 hour, heat stress treatment at 24 hours showed a sum of 1438 genes (heat susceptible) and 1599 genes (heat tolerant) with enhanced regulation and 1237 genes (heat susceptible) and 1754 (heat tolerant) with inhibited regulation. Further analysis of heat responsive genes between heat susceptible and heat tolerant revealed that, among differentially expressed genes (DEGs), HSP family was dominant in heat tolerant. Rice HSFs play a vital role in the regulation of heat stress response by controlling HSPs (Guo et al. 2008). The interplay between HSP101 and HSA32 can affect thermotolerance in rice seedlings (Lin et al. 2014). Overexpression of *Arabidopsis* molecular chaperone HSP101 or the rice TF OsWRKY11 in transgenic rice resulted in decreased leaf necrosis after heat stress treatment (Katiyar-Agarwal et al. 2003; Wu et al. 2009). HSFs can be explored into genome-wide association studies for the identification of heat-tolerant quantitative trait loci (QTLs), which will provide a strong basis for breeding heat-tolerant rice cultivars (Lafarge et al. 2017). Heat stress also yields a unique MADs-box TFs which determines seed size and thermal sensitivity in rice. MAD box TFs and several other genes identified in response to heat stress could be utilized for improving thermotolerant rice cultivars specifically during reproductive development (Chen et al. 2016). Heat stress causes severe reduction of grain weight and quality. Although transcriptome profiling has led to the identification of several genes in response to heat stress, to elucidate the effects of high temperature on the various metabolic and physiological processes of filling grains, it is necessary to understand the crosstalk during stress with the developmental stage of seed ripening or metabolic regulation when the caryopsis is maturing (Tayade et al. 2018). Understanding the functions of various genes and their characterization will be helpful for the development of heat stress tolerant rice varieties with improved morphological, biochemical, and molecular characteristics.

6.1.1.2 Wheat (*Triticum aestivum* L.)

Globally hexaploid wheat (*Triticum aestivum* L.) is the largest contributor of global grain production (Sharma et al. 2015). Increasing temperatures, however, significantly reduce wheat productivity in many regions of the world (Fontana et al. 2015; Hansen et al. 2012; Mueller et al. 2015). As reviewed by Akter and Rafiqul Islam (2017), heat stress alters seed germination and seedling growth, cell turgidity, and plant water-use efficiency. It not only disturbs cellular functions but also generates excessive reactive oxygen species which leads to oxidative stress. Moreover, leaf senescence, reduction of photosynthesis, deactivation of photosynthetic enzymes, and generation of oxidative damages to the chloroplasts are considered the most important responses. These series of events eventually affect reproductive structures and further leads to significant losses in yield. High temperature stress especially during reproductive stages (flowering and grain filling) leads to reduction in kernel weight and number and detrimentally affects yield in wheat (Hays et al. 2007; Maestri et al. 2002; Plaut et al. 2004). Therefore, for a better management of heat stress, the understanding and detailed investigation of heat-tolerance features at molecular levels is necessary. Transcriptome profiling from leaves of wheat seedlings exposed to heat stress (40 °C for 1 hour and 6 hours) alone and in combination with other stresses under controlled conditions showed up- and down-regulation of several genes and TFs (Z. Liu et al. 2015). Furthermore, these studies discovered that the response to wheat seedlings to heat and other stresses was regulated by a crosstalk between *HSFs* and *DREBs*. Yet in another study, heat stress responsive transcriptome analysis from heat susceptible and heat tolerant wheat genotypes was performed using wheat genome array (Qin et al. 2008). Heat stress treated leaves showed high induction of HSPs and other chaperones. More specifically a total of 18 probe sets encoding peptidyl prolyl cis-trans isomerase from chaperone family were up-regulated by heat treatment. Apart from HSPs, heat stress caused changes in expression levels of several genes in leaves of 10-day-old wheat seedlings encoding TFs and various processes such as proteins involved in phytohormone biosynthesis/signaling, calcium, and sugar signal pathways, RNA metabolism, ribosomal proteins, primary and secondary metabolisms, as well as proteins related to other stresses. Differential expression of 313 probe sets noted could be responsible for the difference in heat tolerance capacities of susceptible and tolerant genotypes. It is necessary to develop the transcript database of wheat so that the information could be useful in accelerating the progress of wheat stress genomics as well as breeding programs. RNA sequencing analysis during the anthesis stage in a heat-sensitive wheat *cv.* HD2329 exposed to heat stress generated a wheat transcript dataset for *de novo* assembly (Kumar et al. 2015). Vishwakarma et al. (2018) constructed suppression subtractive hybridization (SSH) library of plants at 93 days old during grain filling stage from leaf samples in a heat stress tolerant Indian bread wheat cultivar C306. Identification of heat stress tolerant gene *TaHsp90* and its heterologous expression validation performed in *Escherichia coli* BL21 can be utilized for developing heat-tolerant transgenic wheat.

6.1.1.3 Maize (*Zea mays* L.)

Maize (*Zea mays* L.) is the most important crop after rice and wheat. Heat stress adversely affects maize, specifically the flowering, pollination, and grain filling stages, which are more susceptible, and ultimately results in the decline of maize production (Zhang et al. 2006; Tao et al. 2016). Although extensive studies have been carried out to understand physiological mechanisms underlying heat stress and heat-shock proteins have been identified, studies on the transcriptome profiling in response to heat stress in different maize varieties/genotypes are meager (Shi et al. 2017). DEGs, in response to heat stress, have been identified in maize, and their functional annotations suggest that the stress response is mediated by various stress responsive pathways involving hormone metabolism and signaling, TFs, very-long-chain fatty acid biosynthesis, and lipid signaling.

In response to heat stress, three pathways – namely carbon metabolism, starch and sucrose metabolism, and carbon fixation – were significantly enriched in maize seedling leaves. Moreover, significant up-regulation of the expression of genes involved in abscisic acid biosynthesis (β-carotene hydroxylase, 9-*cis*-epoxycarotenoid dioxygenase, abscisic acid, and *Arabidopsis* aldehyde oxidase) has also been noted (Li et al. 2017). The role of TF in alleviating the negative effects of high temperature has been demonstrated. Improved plant growth and performance under high temperature stress were observed when maize plants were over-expressed with *OsMYB55*. Transgenic maize lines exhibited higher plant biomass and reduced leaf damage when compared to wild type (Casaretto et al. 2016). Comparative transcriptome profiling showed DEGs in the leaves of two maize genotypes, i.e. heat tolerant and heat susceptible when exposed to 42 °C a number of genes were up-regulated and down-regulated in response to heat stress. DEGs were then subjected to pathway analysis to identify potential candidate genes involved in heat stress resistance. Whereas up-regulated genes were found to be involved in 36 different metabolic pathways (biosynthesis of secondary metabolites, flavonoid biosynthesis, fatty acid biosynthesis, photosynthesis, etc.), down-regulated genes were mainly involved in 23 different metabolic pathways (phenylpropanoid biosynthesis, spliceosome, cytosolic DNA-sensing pathway, glutathione metabolism, etc.). These comparative transcriptomic studies provided new insights into molecular mechanisms involved in heat tolerant and heat susceptible maize genotypes (Shi et al. 2017). Identified candidates can be used in maize-breeding programs to develop superior genotypes. However, heat stress is a complex quantitative trait in maize. Therefore, for successive breeding and effective selection for heat-stress tolerance, replicated field trials in multiple environments are required (Bai 2003). Kandel et al. (2018) suggest that, field experiments should be conducted to identify various biochemical and molecular approaches along with the agronomic management practices which are needed to further investigate the actual high temperature stress responses and their effects on final crop yield.

6.1.1.4 Switchgrass (*Panicum virgatum* L.) and Ryegrass (*Lolium perenne* L.)

Transcriptome analysis in response to heat stress have been performed from switchgrass and perennial ryegrass. To understand the molecular responses to heat stress, Li et al. (2013) performed transcriptome analysis in switchgrass (*Panicum virgatum* L.) Alamo cultivar, a model herbaceous bioenergy crop. An experiment was conducted in a growth chamber, and plants after 30 days of sowing were exposed to long-term heat stress treatment (38°/30 °C, day/night, for 50 days). As a results of the heat stress treatment, switchgrass exhibited a significant reduction in plant height as well as total biomass when compared to the control group. Transcriptome analysis was conducted from heat-stressed and control leaves of switchgrass Affymetrix gene chips. Results revealed that in response to heat stress, among the 5164 differentially expressed probesets, 2076 probesets corresponding to 2002 switchgrass unigenes were up-regulated, and 3088 probesets corresponding to 2809 switchgrass unigenes were downregulated. As switchgrass genes were not well annotated, in gene ontology analysis, rice orthologs (showed 98–100% homology with switchgrass probesets) of the differently regulated transcripts were identified. These results found in response to heat a total of 1158 unique rice orthologs, representing 1478 switchgrass probesets which were induced, and 1857 unique rice orthologs, representing 2584 switchgrass probesets, were repressed. RT-PCR was conducted to validate the transcriptional pattern of differentially expressed 42 randomly selected genes. As switchgrass breeding for increasing biomass and conversion from cellulosic feedstock to ethanol have become a high priority (Casler et al. 2011; Marra et al. 2013), it is essentially important to identify the candidate genes to establish the connections with physiological responses and heat response phenotype in switchgrass. Studies further conclude that genes identified in response to heat stress can be used as biomarkers for germplasm screening for thermotolerance as well as developing heat stress tolerant new plant types (Li et al. 2013). The studies also identified NAC TFs, the largest and plant-specific TF family, which play an important role abiotic stress (He et al. 2005; Mitsuda et al. 2007; Tran et al. 2004). Comprehensive genome-wide analysis in switchgrass identified NAC domain TFs and exploited expanded NAC family with a total of 251 family genes. These studies concluded that the identification of a large number of candidate *NAC* genes can be helpful for future genetic studies on switchgrass (Yan et al. 2017a). Several HSPs have also been identified in switchgrass. Therefore, it can be speculated that understanding the role of TFs will help in the molecular breeding of switchgrass for the improvement of biomass yield under harsh environmental conditions (Yan et al. 2017b).

Owing to its rapid establishment, low growing season as well as high yield characteristics, perennial ryegrass (*Lolium perenne* L.) has become one of the most widely used forage and turf grasses in the world. However, high temperature in warmer and transition regions is a major factor limiting its agronomic performance (Huang et al. 2014). Studies to understand the molecular responses of ryegrass in response to heat stress are meager. Y. Wang et al. (2016) revealed that *HSF* genes are important regulators of heat stress response in perennial ryegrass. RNA sequencing categorized *HSFs* into three subfamilies targeting several other genes such as *HSP*, ascorbate peroxidase (*APX*), inositol-3-phosphate synthase (*IPS*), and galactinol synthase (*GOLS1*). Wang et al. (2013) generated short-term heat stress responsive *de novo* transcriptome using a cDNA library constructed from perennial ryegrass leaves. Information on genes involved in transcriptional regulation, antioxidant responses, plant hormones, signal transduction, and cellular metabolism was obtained which further identified several DEGs including HSFs, HSPs, and antioxidant related genes. RNA-sequencing generated a total of 73 125 unigenes with a total of 20 183 DEGs, including 15 036 up-regulated and 5147 down-regulated DEGs (Wang et al. 2016a), whereas transcriptome assembly of heat stressed ryegrass resulted in 185 671 unigenes (290 039 transcripts) and 121 271 unigenes (65.32% of all unigenes) with a total of 20 198 DEGs consisting of 10 644 induced and 9554 repressed DEGs (Wang et al. 2017). Genes identified in these studies can further provide a basic for future functional characterization of genes and genetic engineering for the improvement of heat-stress tolerance in grasses and other crops.

6.1.1.5 Spinach (*Spinacia oleracea* L.)

To breed heat-tolerant crops, it is essential to understand the molecular mechanisms of crop plants underlying heat stress (Craita and Tom 2013). In view of this, Yan et al. (2013) performed *de novo* transcriptome sequencing and gene expression profiling of an economically and nutritionally important crop: spinach (*Spinacia oleracea* L.). It's a green leafy vegetable and prominently cold tolerant, it possesses heat-sensitive characteristics, and high temperatures affect spinach plant growth, which further results in a significant decrease in yield and quality (Zhao et al. 2018). The molecular information on response of spinach to heat stress is lacking. So far, only a few heat-stress-responsive genes and proteins such as glycine-betaine biosynthesis-related genes, choline monooxygenase, and betaine aldehyde dehydrogenase in spinach have been functionally characterized at the molecular level. For instance, the overexpression of choline monooxygenase and betaine aldehyde dehydrogenase genes have shown an enhanced heat tolerance in transgenic spinach plants (Shirasawa et al. 2006, Yang et al. 2005). Spinach seedlings were cultured in a growth chamber (grown at 24 °C for control), and exposed to 35 °C for 30 minutes (treatment S1) and 5 hours (treatment S2). Transcriptome analysis from spinach leaves revealed that the DEGs from controls vs. treatment S1 was 986, the DEG of control vs. treatment S2 was 1741 and the DEG of treatment S1 vs. treatment S2 was 1587. Comparative transcriptome analysis of spinach leaves identified 896 unique genes in response to heat. DEGs specifically involved in salt stress, organic acid metabolic, and carotenoid metabolic were specifically helpful in understanding the molecular adaptation mechanism of spinach during the heat stress (Yan et al. 2016). These studies can be further explored and utilized to understand the molecular basis of heat-stress tolerance mechanisms in economically important leafy vegetables. In proteomics and phosphoproteomics studies, heat-tolerant spinach variety Sp75 showed accumulation of multiple osmolytes/antioxidants and triggered the diverse ROS-scavenging pathways in response to heat stress. Moreover, it is noteworthy that, in addition to seven proteins which were involved in Ran GTPase-mediated nuclear trafficking, novel 15 heat-increased transcriptional regulation-related proteins were discovered and proposed to be novel heat-responsive factors in spinach. These results provided novel clues toward discovering new heat perception molecules and nodes of fine-tuned heat adaptation networks. Therefore, studies of molecular genetic screens and protein structure-function analysis unraveling the heat-stress-responsive molecular mechanisms in spinach can pave new pathways toward molecular breeding of heat-tolerant cultivars (Zhao et al. 2018).

6.1.1.6 *Brassica rapa* L.

Global gene-expression profiling from B*rassica rapa* inbred lines differing in their thermotolerance capacities identified DEGs which are critical for the heat stress response (Dong et al. 2015). Both the Chinese cabbage lines exhibited induction of HSPs and HSFs along with the up-regulation of ROS-scavenging genes like glutathione peroxidase 2 (*BrGPX2*,

Bra022853), protein kinases, and phosphatases. The data generated in this study could also provide a foundation for molecular marker development for heat-stress tolerance and genetically engineered heat stress tolerant *B. rapa*. Likewise, transcriptome studies in Chinese cabbage (*Brassica rapa* ssp. chinensis) from heat-sensitive and heat-tolerant varieties identified 625 DEGs including various TFs, kinases/phosphatases, genes related to photosynthesis, and effectors of homeostasis. Genes related to photosynthesis such as two photosystem II (PSII) subunit PSBTN (Bra023909) and PSBX (Bra035533) were highly expressed in the heat-tolerant variety "XK" along with one photosystem I subunit D-1 (Bra036240) was also induced in this variety. qRT-PCR validation of selected genes revealed that oxalate oxidase and HSPB27 increased during the heat stress treatment (A. Wang et al. 2016). These studies can be further explored and utilized to understand molecular mechanism of heat-stress tolerance in other *Brassica* species, such as turnip, mustard, etc.

6.1.1.7 Banana (*Musa acuminate* Colla)

Heat stress is considered the major abiotic stress affecting banana (*Musa acuminate* Colla), a nonmodel staple fresh fruit crop worldwide (Z. Liu et al. 2015). In banana, both the whole genome sequencing as well as transcriptome analysis have already been performed. It further helped in understanding signaling mechanism, pathways, metabolomics, and identification of various genes involved in acquiring thermotolerance in banana (D'Hont et al. 2012; Venkataramana et al. 2015; Vidya et al. 2018). RNA sequencing and DEGs were studied from five to six-week-old leaves from tissue cultured banana cv. Grand Naine (*Musa* species AAA Group) seedlings were exposed to four different heat treatments, namely control (without heat treatment), induction stress (gradual increase in temperature from 30 to 42 °C for 2 hours and 30 minutes), induction + lethal (seedlings after induction were shifted to lethal stress 55 °C for 2 hours), and lethal stress (control plants were directly shifted to 55 °C and kept for 2 hours). RNA sequencing analysis revealed a wide range of variability of DEGs in various heat stress treatments. For instance, when compared with control, the number of DEGs under induction treatment was 3936, whereas it was 2268 and 907 in induction + lethal and lethal stress respectively. Gene ontology analysis was performed to investigate the genes which varied in response to heat stress in banana leaf revealed that the majority of alterations were observed in membrane integrity, gene expression (transcription), and protein synthesis categories. HSP family members such as HSPs, HSFs, and small HSPs were mainly altered during induction stress. Among the HSPs including both high and low molecular weight, 37 HSPs were activated during induction, 20 in induction + lethal stress, while only three under lethal stress conditions, revealing their importance in acquiring thermotolerance in banana seedlings. Genome-wide characterization and expression profile analysis of banana seedlings revealed abiotic stress-responsive expressions of HSFs, which may be used as potential candidates for genetic breeding in bananas (Wei et al. 2016). In addition to the HSP family, for acquiring thermotolerance, other stress-associated proteins, ROS-scavenging, fatty acid metabolism, and protein modification were significantly up-regulated during induction stress. Thus this study demonstrated that induction stress (gradual increase in temperature from 30 to 42 °C for 2 hours and 30 minutes) play an important role in cellular and molecular preparedness for higher stress conditions in banana seedlings (Vidya et al. 2018). As protein synthesis and degradation are considered the most sensitive processes to heat stress (Huang and Xu 2008), their functional characterization may help to elucidate their role in plant adaptation to thermotolerance.

6.2 Conclusions

Considering the global scenario of climate change and threat of increasing temperatures, it is critical to develop crop varieties to withstand such conditions. High temperature significantly influences plant growth and development, and the reproductive stages are the most sensitive to heat stress. For this not only a thorough understanding of molecular elements is required but strong efforts are also needed. Advances in high-throughput and next generation sequencing open up opportunities to identify candidate genes responsible for heat-stress tolerance.

Candidate genes provide genetic resources that can be used for developing transgenic plants for heat-stress tolerance. Moreover, molecular regions responsible for heat stress can be used for the development of markers in heat stress breeding programs. So far several transcriptome databases have been established for important crop plants in public domains. Transcriptome studies can be combined with metabolomics to correlate the levels of metabolites produced in response to heat stress with specific candidate genes. Several TF family genes are either up-regulated or repressed under high temperature stress. Genetic engineering approaches prove overexpression of HSPs and other genes are useful in developing heat stress tolerant crops. Thus, the development of crop varieties with increased productivity along with improved quality and yield potential with greater heat-stress tolerance can be achieved. In future, more transcriptomic studies are needed to expedite the development of elite heat stress resistant crop varieties.

References

Akter, N. and Rafiqul Islam, M. (2017). *Agronomy for Sustainable Development* 37: 37. https://doi.org/10.1007/s13593-017-0443-9.

Aprile, A., Mastrangelo, A.M., De Leonardis, A.M. et al. (2009). Transcriptional profiling in response to terminal drought stress reveals differential responses along the wheat genome. *BMC Genomics* 10: 279.

Bai J. (2003) Genetic variation of heat tolerance and correlation with other agronomic traits in a maize (*Zea mays* L.) recombinant inbred line population. Doctoral dissertation. Texas Tech University.

Bheemanahalli, R., Sathishraj, R., Tack, J. et al. (2016). Temperature thresholds for spikelet sterility and associated warming impacts for sub-tropical rice. *Agricultural and Forest Meteorology* 221: 122–130.

Casaretto, J.A., El-Kereamy, A., Zeng, B. et al. (2016). Expression of OsMYB55 in maize activates stress-responsive genes and enhances heat and drought tolerance. *BMC Genomics* 17: 312. https://doi.org/10.1186/s12864-016-2659-5.

Casler, M.D., Tobias, C.M., Kaeppler, S.M. et al. (2011). The switchgrass genome: tools and strategies. *Plant Genome* 4: 273–282.

Chen, S. and Li, H. (2017). Heat stress regulates the expression of genes at transcriptional and post-transcriptional levels, revealed by RNA-seq in *Brachypodium distachyon*. *Frontiers in Plant Science* 7: 2067. https://doi.org/10.3389/fpls.2016.02067.

Chen, C., Begcy, K., Liu, K. et al. (2016). Heat stress yields a unique MADS box transcription factor in determining seed size and thermal sensitivity. *Plant Physiology* 171 (1): 606–622.

Christensen, J.H. and Christensen, O.B. (2007). A summary of the PRUDENCE model projections of changes in European climate by the end of this century. *Climatic Change* 81: 7–30. https://doi.org/10.1007/s10584-006-9210-7.

Cottee, N.S., Wilson, I.W., Tan, D.K., and Bange, M.P. (2014). Understanding the molecular events underpinning cultivar differences in the physiological performance and heat tolerance of cotton (*Gossypium hirsutum*). *Functional Plant Biology* 41: 56–67. https://doi.org/10.1071/F.P.13140.

Craita, E.B. and Tom, G. (2013). Plant tolerance to high temperature in a changing environment: scientific fundamentals and production of heat stress-tolerant crops. *Frontiers in Plant Science* 4: 273.

D'Hont, A. et al. (2012). The banana (*Musa acuminata*) genome and the evolution of monocotyledonous plants. *Nature* 488 (7410): 213–217.

Dong, X., Yi, H., Lee, J. et al. (2015). Global gene-expression analysis to identify differentially expressed genes critical for the heat stress response in *Brassica rapa*. *PLoS One* 10 (6): e0130451. https://doi.org/10.1371/journal.pone.0130451.

Driedonks, N., Xu, J., Peters, J.L. et al. (2015). Multi-level interactions between heat shock factors, heat shock proteins, and the redox system regulate acclimation to heat. *Frontiers in Plant Science* 6: 999. https://doi.org/10.3389/fpls.2015.00999.

Fang, C., Dou, L., Liu, Y. et al. (2018). Heat stress-responsive transcriptome analysis in heat susceptible and tolerant rice by high-throughput sequencing. *Ecological Genetics and Genomics* 6: 33–40.

Fontana, G., Toreti, A., Ceglar, A., and De Sanctis, G. (2015). Early heat waves over Italy and their impacts on durum wheat yields. *Natural Hazards and Earth System Sciences* 15: 1631–1637.

Galsurker, O., Doron-Faigenboim, A., Teper-Bamnolker, P. et al. (2018). Differential response to heat stress in outer and inner onion bulb scales. *Journal of Experimental Botany* 69 (16): 4047–4064.

García-Ortega, L.F. and Martínez, O. (2015). How many genes are expressed in a transcriptome? Estimation and results for RNA-Seq. *PLoS One* 10 (6): e0130262. https://doi.org/10.1371/journal.pone.0130262.

González-Schain, N., Ludovico, D., Lovely, M.F. et al. (2016). Genome-wide transcriptome analysis during anthesis reveals new insights into the molecular basis of heat stress responses in tolerant and sensitive rice varieties. *Plant and Cell Physiology* 57: 57–68.

Guo, J., Wu, J., Ji, Q. et al. (2008). Genome-wide analysis of heat shock transcription factor families in rice and *Arabidopsis*. *Journal of Genetics and Genomics* 35: 105–118.

Hansen, J., Sato, M., and Ruedy, R. (2012). Perception of climate change. *Proceedings of the National Academy of Sciences* 109: 14726–14727.

Hasanuzzaman, M., Nahar, K., Alam, M.M. et al. (2013a). Physiological, biochemical, and molecular mechanisms of heat stress tolerance in plants. *International Journal of Molecular Sciences* 14 (5): 9643–9684. https://doi.org/10.3390/ijms14059643.

Hasanuzzaman, M., Nahar, K., Fujita, M. et al. (2013b). Enhancing plant productivity under salt stress-relevance of poly-omics. In: *Salt Stress in Plants: Omics, Signaling and Responses* (eds. P. Ahmad, M.M. Azooz and M.N.V. Prasad), 113–156. Berlin: Springer.

Hays, D., Mason, E., Do, J.H. et al. (2007). Expression quantitative trait loci mapping heat tolerance during reproductive development in wheat (*Triticum aestivum*). *Wheat Production in Stressed Environments* 12: 373–382.

He, X.J., Mu, R.L., Cao, W.H. et al. (2005). AtNAC2, a transcription factor downstream of ethylene and auxin signaling pathways, is involved in salt stress response and lateral root development. *Plant Journal* 44 (6): 903–916.

Huang, B. and Xu, C. (2008). Identification and characterization of proteins associated with plant tolerance to heat stress. *Journal of Integrative Plant Biology* 50: 1230–1237.

Huang, B., DaCosta, M., and Jiang, Y. (2014). Research advances in mechanisms of turfgrass tolerance to abiotic stresses: from physiology to molecular biology. *Critical Reviews in Plant Sciences* 33: 141–189.

Ishimaru, T., Xaiyalath, S., Nallathambi, J. et al. (2015). Quantifying rice spikelet sterility in potential heat-vulnerable regions: field surveys in Laos and southern India. *Field Crops Research* 190, April 2016: 3–9.

Jagadish, S.V.K., Muthurajan, R., Oane, R. et al. (2010). Physiological and proteomic approaches to address heat tolerance during anthesis in rice (*Oryza sativa* L.). *Journal of Experimental Botany* 61: 143–156.

Jung, K.H., Ko, H.J., Nguyen, M.X. et al. (2012). Genome-wide identification and analysis of early heat stress responsive genes in rice. *Plant Biology* 55: 458.

Kandel, M., Ghimire, S.K., and Shrestha, J. (2018). Mechanisms of heat stress tolerance in maize. *Azarian Journal of Agriculture* 5 (1): 20–27.

Katiyar-Agarwal, S., Agarwal, M., and Grover, A. (2003). Heat-tolerant basmati rice engineered by over-expression of hsp101. *Plant Molecular Biology* 51: 677–686.

Kumar, R.R., Goswami, S., Sharma, S.K. et al. (2015). Harnessing next generation sequencing in climate change: RNA-Seq analysis of heat stress-responsive genes in wheat (*Triticum aestivum* L.). *Omics: A Journal of Integrative Biology* 19 (10): 632–647.

Lafarge, T., Bueno, C., Frouin, J. et al. (2017). Genome-wide association analysis for heat tolerance at flowering detected a large set of genes involved in adaptation to thermal and other stresses. *PLoS One* 12: e0171254.

Li, Y.F., Wang, Y., Tang, Y. et al. (2013). Transcriptome analysis of heat stress response in switchgrass (*Panicum virgatum* L.). *BMC Plant Biology* 13: 153. https://doi.org/10.1186/1471-2229-13-153.

Li, T., Xu, X., Li, Y. et al. (2015). Comparative transcriptome analysis reveals differential transcription in heat-susceptible and heat-tolerant pepper (*Capsicum annum* L.) cultivars under heat stress. *Journal of Plant Biology* 58: 411–424.

Li, P., Cao, W., Fang, H. et al. (2017). Transcriptomic profiling of the maize (*Zea mays* L.) leaf response to abiotic stresses at the seedling stage. *Frontiers in Plant Science* 8: 290. https://doi.org/10.3389/fpls.2017.00290.

Lin, M.Y., Chai, K.H., Ko, S.S. et al. (2014). A positive feedback loop between heat shock protein 101 and heat stress-associated 32-KD protein modulates long-term acquired thermotolerance illustrating diverse heat stress responses in rice varieties. *Plant Physiology* 164: 2045–2053.

Liu, Z., Xin, M., Qin, J. et al. (2015). Temporal transcriptome profiling reveals expression partitioning of homeologous genes contributing to heat and drought acclimation in wheat (*Triticum aestivum* L.). *BMC Plant Biology* 15: 152. https://doi.org/10.1186/s12870-015-0511-8.

Lowe, R., Shirley, N., Bleackley, M. et al. (2017). Transcriptomics technologies. *PLoS Computational Biology* 13 (5): e1005457. https://doi.org/10.1371/journal.pcbi.1005457.

Maestri, E., Klueva, N., Perrota, C. et al. (2002). Molecular genetics of heat tolerance and heat shock proteins in cereals. *Plant Molecular Biology* 48: 667–681.

Marra, M., Keene, T., Skousen, J., and Griggs, T. (2013). Switchgrass yield on reclaimed surface mines for bioenergy production. *Journal of Environmental Quality* 42: 696–703.

Mitsuda, N., Iwase, A., Yamamoto, H. et al. (2007). NAC transcription factors, NST1 and NST3, are key regulators of the formation of secondary walls in woody tissues of *Arabidopsis*. *The Plant Cell* 19 (1): 270–280.

Mochida, K. and Shinozaki, K. (2010). Genomics and bioinformatics resources for crop improvement. *Plant and Cell Physiology* 51 (4): 497–523. https://doi.org/10.1093/pcp/pcq027.

Mueller, B., Hauser, M., Iles, C. et al. (2015). Lengthening of the growing season in wheat and maize producing regions. *Weather and Climate Extremes* 9: 47–56.

Plaut, Z., Butow, B.J., Blumenthal, C.S., and Wrigley, C.W. (2004). Transport of dry matter into developing wheat kernels and its contribution to grain yield under post-anthesis water deficit and elevated temperature. *Field Crops Research* 86: 185–198.

Prasad, P.V.V., Boote, K.J., Allen, L.H. et al. (2006). Species, ecotype and cultivar differences in spikelet fertility and harvest index of rice in response to high temperature stress. *Field Crops Research* 95: 398–411.

Qi, Y., Wang, H., Zou, Y. et al. (2011). Over-expression of mitochondrial heat shock protein 70 suppresses programmed cell death in rice. *FEBS Letters* 585 (1): 231–239.

Qin, D., Wu, H., Peng, H. et al. (2008). Heat stress-responsive transcriptome analysis in heat susceptible and tolerant wheat (*Triticum aestivum* L.) by using wheat genome array. *BMC Genomics* 9: 432.

Sailaja, B., Mangrauthia, S.K., Sarla, N., and Voleti, S.R. (2014). Transcriptomics of heat stress in plants. In: *Improvement of Crops in the Era of Climatic Changes*, vol. 2 (eds. P. Ahmad et al.). New York: Springer Science+Business Media https://doi.org/10.1007/978-1-4614-8824-8_3.

Sánchez-Rodríguez, E., Moreno, D.A., Ferreres, F. et al. (2011). Differential responses of five cherry tomato varieties to water stress: changes on phenolic metabolites and related enzymes. *Phytochemistry* 72 (8): 723–729.

Sathishraj, R., Bheemanahalli, R., Ramachandran, M. et al. (2015). Capturing heat stress induced variability in spikelet sterility using panicle: leaf and air temperature under field conditions. *Field Crops Research* 190: 10–17.

Shah, F., Huang, J., Cui, K. et al. (2011). Impact of high-temperature stress on rice plant and its traits related to tolerance. *Journal of Agricultural Science* 149: 545–556.

Sharma, I., Tyagi, B.S., Singh, G. et al. (2015). Enhancing wheat production: a global perspective. *Indian Journal of Agricultural Sciences* 85: 3–13.

Shi, J., Yan, B., Lou, X. et al. (2017). Comparative transcriptome analysis reveals the transcriptional alterations in heat-resistant and heat-sensitive sweet maize (*Zea mays* L.) varieties under heat stress. *BMC Plant Biology* 17 (1): 26. https://doi.org/10.1186/s12870-017-0973-y.

Shirasawa, K., Takabe, T., Takabe, T., and Kishitani, S. (2006). Accumulation of glycinebetaine in rice plants that overexpress choline monooxygenase from spinach and evaluation of their tolerance to abiotic stress. *Annals of Botany* 98: 565–571.

Siebert, S. and Ewert, F. (2014). Future crop production threatened by extreme heat. *Environmental Research Letters* 9: 041001.

Suzuki, N., Rivero, R.M., Shulaev, V. et al. (2014). Abiotic and biotic stress combinations. *The New Phytologist* 203: 32–43. https://doi.org/10.1111/nph.12797.

Tao, Z.Q., Chen, Y.Q., Li, C. et al. (2016). The causes and impacts for heat stress in spring maize during grain filling in the North China Plain: a review. *Journal of Integrative Agriculture* 15 (12): 2677–2687.

Tardieu, F. and Hammer, G. (2012). Designing crops for new challenges. *European Journal of Agronomy* 42: 1–2.

Tayade, R., Nguyen, T.D., Oh, S.A. et al. (2018). Effective strategies for enhancing tolerance to high-temperature stress in rice during the reproductive and ripening stages. *Plant Breeding and Biotechnology* 6: 1–18.

Thakur, P., Kumar, S., Malik, J.A. et al. (2010). Cold stress effects on reproductive development in grain crops: an overview. *Environmental and Experimental Botany* 67: 429–443. https://doi.org/10.1016/j.envexpbot.2009.09.004.

Tran, L.S., Nakashima, K., Sakuma, Y. et al. (2004). Isolation and functional analysis of *Arabidopsis* stress-inducible NAC transcription factors that bind to a drought-responsive cis-element in the early responsive to dehydration stress 1 promoter. *The Plant Cell* 16 (9): 2481–2498.

Venkataramana, K.R., Sampangi-Ramajah, H.M., Ajitha, R. et al. (2015, 2015). Insights into *Musa balbisiana* and *Musa acuminata* species divergence and development of genic microsatellites by transcriptomics approach. *Plant Gene* 4: 78–82.

Vidya, S.M., Kumar, H.V., Bhatt, R.M. et al. (2018). Transcriptional profiling and genes involved in acquired thermotolerance in banana: a non-model crop. *Scientific Reports* 8: 10683. https://doi.org/10.1038/s41598-018-27820-4.

Vishwakarma, H., Junaid, A., Manjhi, J. et al. (2018). Heat stress transcripts, differential expression, and profiling of heat stress tolerant gene TaHsp90 in Indian wheat (*Triticum aestivum* L.) cv C306. *PLoS One* 13 (6): e0198293. https://doi.org/10.1371/journal.pone.0198293.

Wan, X.L., Zhou, Q., Wang, Y.Y. et al. (2015). Identification of heat-responsive genes in carnation (*Dianthus caryophyllus* L.) by RNA-seq. *Frontiers in Plant Science* 6: 519. https://doi.org/10.3389/fpls.2015.00519.

Wang, Y., Dai, Y., Tao, X. et al. (2016). Heat shock factor genes of tall fescue and perennial ryegrass in response to temperature stress by RNA-Seq analysis. *Frontiers in Plant Science* 6: 1226. https://doi.org/10.3389/fpls.2015.01226.

Wang, K., Liu, Y., Tian, J. et al. (2017). Transcriptional profiling and identification of heat-responsive genes in perennial ryegrass by RNA-sequencing. *Frontiers in Plant Science* 8: 1032.

Wei, Y., Hu, W., Xia, F. et al. (2016). Heat shock transcription factors in banana: genome-wide characterization and expression profile analysis during development and stress response. *Scientific Reports* 6: 36864. https://doi.org/10.1038/srep36864.

Wu, X., Shiroto, Y., Kishitani, S. et al. (2009). Enhanced heat and drought tolerance in transgenic rice seedlings overexpressing OsWRKY11 under the control of HSP101 promoter. *Plant Cell Reports* 28: 21–30.

Yan, J. et al. (2016). *De novo* transcriptome sequencing and gene expression profiling of spinach (*Spinacia oleracea* L.) leaves under heat stress. *Scientific Reports* 6: 19473.

Yan, H., Zhang, A., Ye, Y. et al. (2017a). Genome-wide survey of switchgrass NACs family provides new insights into motif and structure arrangements and reveals stress-related and tissue-specific NACs. *Scientific reports* 7 (1): 3056. https://doi.org/10.1038/s41598-017-03435-z.

Yan, H., Zhang, A., Chen, J. et al. (2017b). Genome-wide analysis of the PvHsp20 family in switchgrass: motif, genomic organization, and identification of stress or developmental-related Hsp20s. *Frontiers in plant science* 8: 1024. https://doi.org/10.3389/fpls.2017.01024.

Yang, X., Liang, Z., and Lu, C. (2005). Genetic engineering of the biosynthesis of glycinebetaine enhances photosynthesis against high temperature stress in transgenic tobacco plants. *Plant Physiology* 138: 2299–2309.

Yang, J., Chen, X., Zhu, C. et al. (2015). Using RNA-seq to profile gene expression of spikelet development in response to temperature and nitrogen during meiosis in rice (*Oryza sativa* L.). *PLoS One* 10 (12): e0145532. https://doi.org/10.1371/journal.pone.0145532.

Zhang, B., Dong, S., Hu, C., and Wang, K. (2006). Research progress on heat stress and adaptation of maize. *J Weifang Univ.* 6 (6): 90–94.

Zhao, Q., Chen, W., Bian, J. et al. (2018). Proteomics and phosphoproteomics of heat stress-responsive mechanisms in spinach. *Frontiers in Plant Science* 9: 800. https://doi.org/10.3389/fpls.2018.00800.

7

Proteomics as a Tool for Characterizing the Alteration in Pathways Associated with Defense and Metabolite Synthesis

Reetika Mahajan[1] *and Sajad Majeed Zargar*[2]

[1] *Govt. Degree College Poonch, Jammu and Kashmir, India*
[2] *Division of Plant Biotechnology, Sher-e-Kashmir University of Agricultural Science and Technology of Kashmir, Shalimar, Srinagar, Jammu and Kashmir, India*

7.1 Introduction

Technological advances have changed the overall scenario of crop improvement in the present era. Various omics approaches like genomics, transcriptomics, proteomics, metabolomics, and ionomics play important roles in crop improvement. Proteomics is a complex field which deals with the study of structure and function of proteins. Proteomics as an advanced omics tools could be used to understand the mechanism of plant response to different stresses. This branch of omics could also be used to have insight of the pathways of different defense mechanism and metabolite synthesis in plants. At the time of unfavorable conditions like biotic stresses (insect, fungal, bacterial, and viral), plants undergo certain changes in their metabolite synthesis pathways that provide defense to the plant to cope with the stress. These stresses lead to alteration in the plants' efficiency. When encountering biotic stress, plants undergo two different pathways for recognizing and resisting pathogen attack. The first pathway deals with DAMPs (danger-associated molecular patterns) and PAMPs (pathogen-associated molecular patterns) formation and in the second pathway ETI (effector-triggered immunity) and PTI

Heat Stress Tolerance in Plants: Physiological, Molecular and Genetic Perspectives, First Edition.
Edited by Shabir Hussain Wani and Vinay Kumar.

(PAMPs-triggered immunity) are formed after the recognition of pathogen by specific pathogen effector molecules (De Wit 2007; Dodds and Rathjen 2010; Lodha and Basak 2012). In this chapter, we discuss different proteomics technologies which could be used as alternatives to characterize the alteration in pathways associated with defense and metabolite synthesis.

7.2 What Is Proteomics?

The final product of central dogma of life is protein and the total number of proteins expressed in a cell is called a proteome. The term "proteome" can also be define as the entire PROTEin complement expressed by a genOME (Wilkins et al. 1996). The term "proteomics" was coined back in 1996 by Wilkins and colleagues. Proteomics means the study of proteome structurally and functionally in a cell, tissue, or whole organism at a given time under defined conditions (Wilkins et al. 1996). Proteomics is considered the next and advanced step to genomics and transcriptomics as it is more complicated, accurate, sensitive, with high speed and throughput mass spectrometry (MS), and uses powerful analytical software. Advantages of proteomics over genomics have given this approach more attention, as, by using this approach, one can directly find out the identity and amount of each and every protein present in a cell and its function, which can mediate specific cellular processes (Nat et al. 2007). There are two branches of proteomics: structural (deals with the identification of structure of protein, i.e. primary, secondary, and tertiary structures) and functional (deals with the methods which are used to identify protein functions). For the first time, proteomics research was carried out on yeast and mammalian cell lines (Schulze and Usadel 2010). It was found that plant proteomics research is far behind the human and animal proteomics. However, during the last two decades there has been an increase in plant proteomics research. Now, most of the proteomics research in plants is carried out in model organisms and cash crops, e.g. *Arabidopsis* and rice (Abdallah et al. 2012). Techniques used for quantitative proteomics analysis are mostly the same for both animals and plants. These techniques have given a boost to plant proteome studies, which has resulted in conducting research on differential expression, as well as global and organellar proteomic elucidation. However, in plants proteomic sample handling is very crucial as plants are different to animals in terms of genome, physiology, and culture practices.

7.3 Need of Proteomics in Post-genomic Era

The present era, referred to as the post-genomic era, includes the advance technologies like transcriptomics, proteomics, ionomics, and metabolomics. The genome of an organism often changes, whereas the proteome of one cell differs from other cells, owing to change in biochemical interaction of the protein with the genome and environment under different stresses. Post-transcriptional and translation modifications result in expression of functional and nonfunctional proteins in a particular cell. Thus proteomics is needed to have accurate knowledge of the protein present in a cell (Gygi et al. 1999; Duley and Grover 2001; Park 2004; Thurston et al. 2005). Proteomics is also needed to unravel the function of genes identified by genomics and transcriptomics (Patterson and Aebersold 2003). Modern proteomics techniques could be used to perform comparative study of the protein content of cells under stress conditions.

7.4 Different Branches of Proteomics

"Proteomics" refers to the study of structure and function of a particular protein. Thus this field of modern science has many branches (Figure 7.1):

- *Protein separation*
  This is the first and most important branch of proteomics. This deals with the isolation of protein from the cell or organelle. The most commonly used methods for protein isolation in plants are the TCA/acetone method and the phenol method.

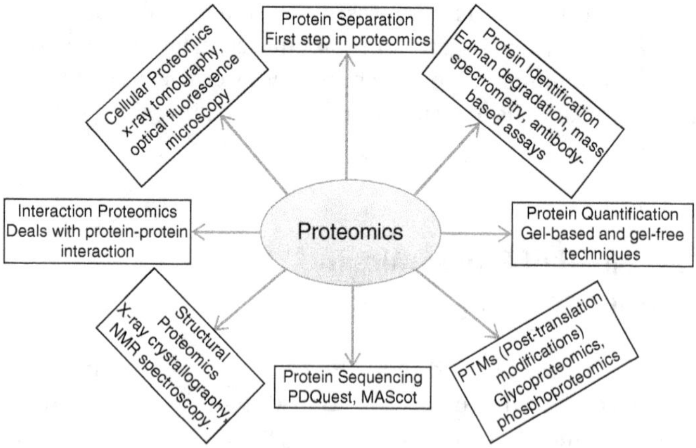

Figure 7.1 Different branches of proteomics used in modern research with the techniques utilized.

- *Protein identification*
  After protein separation, the identification of separated protein is important. Different methods are used to identify proteins. This includes sequencing of proteins by Edman degradation (low-throughput method), identification of proteins based on MS (high-throughput method), and antibody-based assays (sequence motif specific method).
- *Protein quantification*
  Identified proteins need to be quantified for further analysis. Quantification of proteins can be done by two methods: gel-based and gel-free methods. Under gel-based methods, techniques like SDS PAGE, 2D-PAGE, and 2D-DIGE are used. Gel-free methods include various tagging or chemical modification methods, such as isotope-coded affinity tags (ICATs) and ITRAQ (isobaric tagged for relative and absolute quantitation).
- *Post-translation modifications (PTMs)*
  This branch of proteomics deals with modifications of proteins after translation. This includes certain special methods like phosphoproteomics and glycoproteomics, which help

in studying PTMs, like phosphorylation and glycosylation, respectively.

- *Protein sequencing*
  This branch of proteomics mainly deals with bioinformatic-like database searches for possible protein matches, prediction of protein function from identified sequences, functional assignment of domains, and evolutionary relationship of proteins.
- *Structural proteomics*
  This deals with the determination of identified protein structure in 3D space. X-ray crystallography and nuclear magnetic resonance (NMR) spectroscopy are the two methods used in structure identification.
- *Interaction proteomics*
  It deals with the protein–protein interaction at the atomic, molecular, and cellular levels. This includes techniques like affinity purification, Y2H (yeast two-hybrid), TAP (tandem affinity purification), and *in silico* methods including structure and sequence-based approaches.
- *Cellular proteomics*
  This branch deals with the location of proteins and protein–protein interactions in whole cells during key cell events. X-ray tomography and optical fluorescence microscopy are the most commonly used techniques in cellular proteomics.

7.5 Techniques Used in Quantitative Proteomics

Quantitative proteomics deals with the measurement of protein expression under a particular condition. This approach can be used to identify differentially expressed proteins among two contrasting characteristics, like treated vs. control samples. Thus this approach can be considered a relative quantification of expressed proteins. Quantitative proteomics can be divided into two categories: (i) gel-based or gel-free methods and (ii) label-based or label-free methods (Figure 7.2). An overview

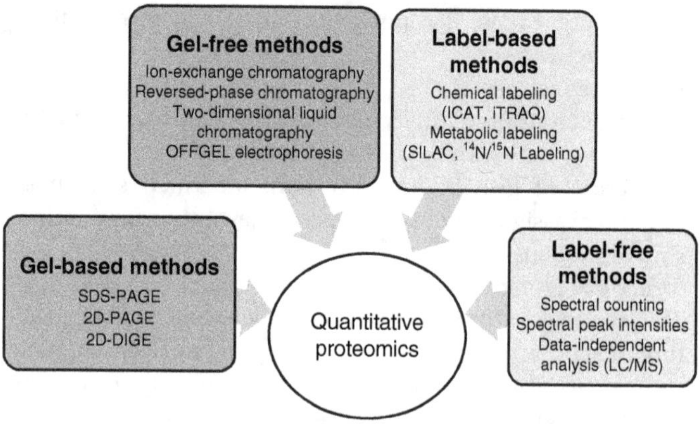

Figure 7.2 Showing two different categories of quantitative proteomics. Quantitative proteomics is categorized into gel-based or gel-free and label-based or label-free methods.

of the methods commonly used in plant proteomics is given in the next section.

7.5.1 Gel-based and Gel-free Methods

Here gel-based method includes SDS PAGE (sodium dodecyl sulfate polyacrylamide gel electrophoresis), 2D-PAGE (two-dimensional polyacrylamide gel electrophoresis), 2D-DIGE (two-dimensional difference gel electrophoresis) techniques, whereas gel-free methods includes ion-exchange chromatography, reversed-phase chromatography, two-dimensional liquid chromatography, and OFFGEL electrophoresis. The most commonly used technique from this category includes gel-based proteomics. Thus under this section only gel-based methods are discussed briefly.

- *SDS PAGE (sodium dodecyl sulfate polyacrylamide gel electrophoresis)*
  SDS-PAGE is one of the traditional gel-based methods based on the molecular weight separation of protein complexes. This method can be used to separate protein under native conditions, which is followed by individual proteins separation under denaturing conditions. The gel is then stained with Coomassie Brilliant Blue G-250 dye to visualize the

different protein bands. This step provides an insight into stoichiometry of protein complexes. Earlier, SDS-PAGE was the most popular technique for protein separation as this technique helps in a faster determination of protein molecular weight and can be used for protein purity assessment. SDS-PAGE can be used for comparative studies of protein, number, and size of polypeptide subunits analysis, can be coupled with western blot for immunodetection, in 2-DE as second dimension for separating proteins on molecular weight bases. But the limitation of this technique is that it cannot separate proteins on the basis of charge, which leads to the development of advanced gel-based techniques. Blue native PAGE (BN-PAGE) is one of the advanced techniques for the separation of native protein complexes present in the electron transfer chain in the plastids and mitochondria of plants (Eubel et al. 2005). Schagger and Von Jagow (1991) used BN-PAGE for studying protein complexes and separating the protein complexes of respiratory chain of mitochondria. BN-PAGE was used for the analysis of wheat chloroplast proteome (Meng et al. 2011).

- *Two-dimensional gel electrophoresis (2-DE)*
  2-DE is an advanced gel-based method which works on isoelectric point (pI) and molecular size separation of proteins from protein mixture (Klose 1975; O'Farrell 1975). In this method, proteins are first separated based on their net charge at different pH by isoelectric focusing (IEF) and then the proteins are separated based on their molecular weight, which is known as "second dimensional separation" (Gorg et al. 2004). For proper protein solubility chaotropes like urea and thiourea (Gordon and Jencks 1963) and zwitterionic detergents like Triton X-100 can be used in 2-DE buffer (Vertommen et al. 2011). Chaotropes can be used to weaken hydrophobic interactions and hydrogen bonds present between proteins so that they can unfold the protein complex (Gordon and Jencks 1963). For proper denaturation, ionic detergents SDS (sodium dodecyl sulfate) can be used in 2-DE buffers. SDS detergents can bind to the hydrophobic amino acid by their hydrophobic hydrocarbon tail to form amino acid–detergent interaction, which helps to promote denaturation. The protein spots separated after 2-DE can be

visualized by staining the gels with different stains like silver, Coomassie, or fluorescent stains like sypro ruby, lava, and deep purple (Patterson and Aebersold 2003; Nat et al. 2007). After visualizing the differentially expressed protein spots, they are excised either robotically or manually followed by analysis by MS (Zhu et al. 2003; Rose et al. 2004) and then are analyzed by using software like PDQuest, Mascot, Melanie, Phoretix, Progenesis, Z3 or Z4000 (Righetti et al. 2004). 2-DE can be also used to study PTMs like phosphoproteomics for phosphorylation and glycoproteomics for glycosylation (Gorg et al. 2004; Wittmann-Liebold et al. 2006). With the development of 2-DE, more than 10 000 protein spots and multiple spots of different molecular forms of the same protein can be visualized on a single gel (Schulze and Usadel 2010). Certain issues like reproducibility issues, low abundant proteins, and highly acidic or basic proteins cannot be visualized properly and complicated automation of technique has limited the utilization of 2-DE (Gygi et al. 2000; Tonge et al. 2001; Ong and Pandey 2001; Lilley et al. 2002).

- *2D-DIGE (two-dimensional difference gel electrophoresis)*
  2D-DIGE is an advanced gel-based proteomics technique which can detect protein at the sub-picomolar level and can be used to study membrane proteins (Vertommen et al. 2011). 2D-DIGE helps to analyze differential protein regulation between two samples (control and target). This technique is based on the covalent binding of NHS-ester reactive group present in the dye to the ε-amino group of protein lysine through amide linkage. In this technique, samples to be analyzed are labeled with one of two different but structurally similar fluorescent CyDyes (Cy2, Cy3, and Cy5) before the electrophoresis step. The dye is added to the samples in accordance with the availability of proteins in the reaction. The two labeled samples are then mixed and run on a single gel where the proteins present in one sample will superimpose with their counterpart protein which is differentially labeled in another sample (Unlu et al. 1997). After electrophoresis, the gel is scanned at two different wavelengths which excite the dye molecules to detect the association of protein spots with dyes (Unlu et al. 1997). The final images of the gel are then imported into 2D-DIGE

image analyzer software, such as DeCyder, Proteomweaver, PDQuest, and Progenesis Same Spots (Marouga et al. 2005; Lodha et al. 2013). The 2D-DIGE method is preferable to 2-DE as it is highly sensitive and reproducible (Van den Bergh and Arckens 2005). 2D-DIGE can also identify protein changes rapidly between two contrasting samples on the same 2D gel, which reduces gel-to-gel variations. 2D-DIGE has been successful in investigating symbiosis and pathogenesis-related protein in *Medicago truncatula* (Van Noorden et al. 2007; Schenkluhn et al. 2010), impact of drought in oak (Sergeant et al. 2011), frost in *Arabidopsis* (Li et al. 2011), ozone, and heavy metals in poplar (Kieffer et al. 2008; Durand et al. 2010; Bohler et al. 2011). The drawbacks of gel-based techniques like these are that they are not sensitive to low abundant proteins, poor reproducibility, and these techniques can only analyze a fraction of the proteome. Thus to overcome these disadvantages, gel-free proteomics has been developed which can analyze low abundant proteins in samples (Zhu et al. 2003; Lambert et al. 2005; Nat et al. 2007).

7.5.2 Label-based and Label-free Methods

Label-based methods are divided into two types of labeling: (i) chemical labeling including proteolytic labeling, ICAT, isotope-coded protein labelling (ICPL), iTRAQ, TandemMass Tag (TMT) and (ii) metabolic labeling (SILAC, 14N/15N labeling) on the other side label-free methods involves spectral counting, spectral peak intensities, data-independent analysis (LC/MS).

- *ICAT (isotope-coded affinity tags)*
  ICAT is one of the gel-free chemical labeled proteomic approaches developed for the identification and quantification of proteins present in two different cells (Gygi et al. 1999; Nat et al. 2007). The first label used for ICAT consists of three functional groups: (i) specific chemical reactive group (binds to sulfhydryl groups of cysteinyl residues), (ii) isotopically coded linker (with light or heavy isotopes), and (iii) biotin tag (for affinity purification) (Gygi et al. 1999). In this method, the two different populations under study were tagged with

different isotopically coded linker (light or heavy isotopes of ^{13}C respectively). After tagging, the populations were pooled and cleaved enzymatically (proteolytically) to generate tagged peptides. These peptides were passed through biotin-avidin affinity chromatography to reduce the complexity of samples before analyzing them by liquid chromatography-electron spray tandem MS (Nat et al. 2007; Gygi et al. 1999). The isotope labeling in this method is done by the pooling of samples before treating the samples with enzyme (protease) to reduce vial-to-vial variations (Thelen and Peck 2007). Quantitative as well as qualitative changes of protein due to biotic and abiotic stresses can be identified by ICAT.

- *iTRAQ (isobaric tagged for relative and absolute quantitation)* iTRAQ is also gel-free chemically labeled proteomics approach with the ability to quantify protein from different sources in a single experiment (Ross et al. 2004; Agarwal et al. 2006; Zieske 2006; Lund et al. 2007). In the iTRAQ method, peptides are labeled with isobaric tags. The isobaric tags are used to label N-termini and side chain amine of digested peptides. After labeling, nano chromatography is performed to separate digested peptides from the mixture and then these peptides are analyzed by MS/MS. In the end, the separated labeled peptides and their corresponding proteins are identified via performing a database search. iTRAQ can quantify and analyze eight samples in one experiment (Ross et al. 2004).

7.6 Role of Proteomics in Studying Alteration in Pathways Associated with Defense and Metabolite Synthesis

In the last two decades, various reports have been published which show the involvement of proteomics techniques in identifying differentially expressed proteins and unrevealing complex mechanism exhibit by plants under stress conditions (Table 7.1). Proteomics has been used to understand the pathways associated with defense and metabolite synthesis in plants. To withstand the stress condition, plants adapt one of

Table 7.1 A summary of important defense-related experiments using proteomics.

Plant species	Pathogen	Proteomic technique used	Reference
Arabidopsis thaliana	*Alternaria brassicicola*	2D-PAGE, LC MS/MS	Mukherjee et al. (2010)
	Spodoptera exigua	2D-PAGE, MALDI-TOF	Zhang et al. (2010)
	Pseudomonas syringe pv. Tomato DC3000	Nitration of proteins, 2D-PAGE LC MS/MS	Cecconi et al. (2009)
	Fusarium elicitor	2D-DIGE, MALDI-TOF MS	Chivasa et al. (2006)
	Pseudomonas syringae	2D-PAGE, LC-MS/MS	Jones et al. (2006)
Brassica napus	*Sclerotinia sclerotiorum*	2D-PAGE, ESI-q-TOF MS/MS	Liang et al. (2008)
Brassica napus and *Brassica carinata*	*Leptosphaeria maculans*	2D-PAGE, ESI-MS	Sharma et al. (2008)
Capsicum annuum	*Capsicum annuum*	2D-PAGE, MALDI-TOF/TOF-MS	Wongpia and Lomthaisong (2010)
Cucumis sativa	*Trichoderma asperellum*	2-DE PAGE, MALDI-TOF/TOF	Segarra et al. (2007)
Coffea canephora	*Meloidogyne paranaensis*	2-DE	Franco et al. (2010)
Citrus sinensis	*Xanthomonas axonopodis* pv. *citri*	2D-PAGE, N-terminal sequencing	Mehta and Rosato (2001)
Capsicum annuum	Tobacco mosaic virus	2D-PAGE, MALDI-TOF MS	Lee et al. (2009)
Fagus sylvatica	*Phytophthora citricola*	SDS-PAGE, 2D-PAGE, ESI, LC-ESI	Valcu et al. (2009)
Glycine max	*Bacillus amyloliquefaciens*	2D PAGE	Buensanteai et al. (2009)
	Xanthomonas axonopodis pv. *Glycine*	2D-PAGE	Buensanteai et al. (2009)
	Heterodera glycines	2D-PAGE, q-TOF MS/MS	Afzal et al. (2009)
Gossypium hirsutum	*Meloidogyne paranaensis*	2D-PAGE	Franco et al. (2010)

(Continued)

Table 7.1 (Continued)

Plant species	Pathogen	Proteomic technique used	Reference
Nicotiana benthamiana	PMMoV-S	2D-PAGE, N-terminal sequencing	Perez-Bueno et al. (2004)
Lysopersicon hirsutum	*Clavibacter michiganensis* sp. *michiganensis*	2D-PAGE, LCQ-Deca ion trap MS	Coaker et al. (2004)
Lysopersicon esculentum	*Fusarium oxysporum*	2D-PAGE, MALDI-TOF MS, LC-QTOF MS/MS	Houterman et al. (2007)
Medicago truncatula	*Aphanomyces euteiches*	2D-PAGE, MALDI, PMF	Colditz et al. (2004)
Medicago truncatula	*Orobanche crenata*	2D- PAGE, PMF, MALDI-TOF	Castillejo et al. (2009)
Nicotinia tabacum	Pepper mild mottle virus	2D-PAGE, N-terminal sequencing	Perez-Bueno et al. (2004)
Oryza sativa	Rice yellow mottle virus (RYMV)	Exclusion chromatography, SDS-PAGE, Nano LC-MS/MS	Delalande et al. (2005)
	RYMV	2D-PAGE, MALDI, LC-MS/MS	Ventelon-Debout et al. (2004)
	Nilaparvata lugens	2D-PAGE, HPLC-MS/MS	Wei et al. (2009)
	Rhizoctonia solani	2D-PAGE, ESI Q-TOF MS, MS/MS	Lee et al. (2009)
	Magnaporthe grisea	2D-PAGE,MALDI-TOF	Kim et al. (2004)
	Xanthomonas oryzae pv. *oryzae*	2D-PAGE, MALDI TOF Edman degradation	Mahmood et al. (2006)

Olea europaea ssp. *Europaea*	*Pseudomonas savastanoi* pv. *Savastanoi*	2D-PAGE, MALDI TOF	Campos et al. (2009)
Passiflora edulis	*Xanthomonas axonopodis* pv. *Passiflorae*	2D- PAGE, Edman sequencing	Tahara et al. (2003)
Prunus persica, and *Prunus serotina*	*Plum pox virus*	SDS-PAGE, IEF, MALDI-TOF	Diaz-Vivancos et al. (2006)
Solanum lycopersicum	*Ralstonia solanacearum*	2D- PAGE, n and cLC *de novo*	Dahal et al. (2009)
Saintpaulia ionantha	*Dickeya dadantii* (syn. *Erwinia chrysanthemi*)	2D-PAGE, MALDI-TOF MS	Babujee et al. (2007)
Triticum aestivum	*Puccinia triticina*	2D-PAGE, MALDI-QqTOF MS/MS	Rampitsch et al. (2006)
	Fusarium graminearum	2D-PAGE, LC-MS/MS	Zhou et al. (2006)
Zea mays	*Fusarium verticillioides*	2D-PAGE, MALDI-TOF-MS, LCQ ion trap MS	Campo et al. (2004)
Vigna mungo	Mung bean yellow mosaic virus	MALDI-TOF/TOF	Kundu et al. (2011)

Source: Lodha et al. 2013.

the two defense mechanisms, i.e. systemic acquired resistance (SAR) or plant-induced systemic resistance (PISR). At the time of stress, an alteration in protein expression present in the plant metabolic pathways provides resistance to plants against the stress. Thus there is a need to explore those alterations via different omics approaches so as to develop resistant plant varieties (Vanderschuren et al. 2013). A proteomic analysis of *Arabidopsis thaliana* recombinant inbred lines (RILs) for studying the effect of insect *Plutella xylostella* revealed significant differences in the proteomes of resistant and susceptible RILs. This study showed that some of the proteins responsive to the insect attack were associated with the alteration in tricarboxylic acid (TCA) cycle and carbon metabolism for providing defense during insect attack (Collins et al. 2010). Proteomic investigation of response to glucosinolate biosynthesis alteration has been carried out in *A. thaliana.* The 2D-DIGE and iTRAQ methods were used to identify proteins involved in metabolism including photosynthesis and hormone metabolism, protein binding, energy, stress, and defense, which showed significant responses to glucosinolate alteration in *A. thaliana.* Proteomics approaches were also used to study redox protein response to methyl jasmonate in *A. thaliana* (Alvarez et al. 2009). 2-DE comparative analysis of response of potato plant to *Leptinotarsa decemlineata* and *Macrosiphum euphorbiae* was carried out in an experiment which resulted in the identification of 500 leaf proteins having differential impacts of wound and insect on defense and photosynthesis-related components of the potato leaf proteome (Duceppe et al. 2012). 2-DE method coupled with MS/MS was used to study programmed cell death and defense signaling by using rice lesion mimic mutant cell death and resistance (cdr). These mutants induce a series of defense responses, including expression of defense-related genes and high accumulation of phytoalexins, indicating that the cdr mutants are useful material to study programmed cell death and defense signaling in rice. A total of 37 differentially expressed defense-related proteins between cdr2 and wild type were identified. It was suggested that the programmed cell death that occurs in the cdr2 mutant was

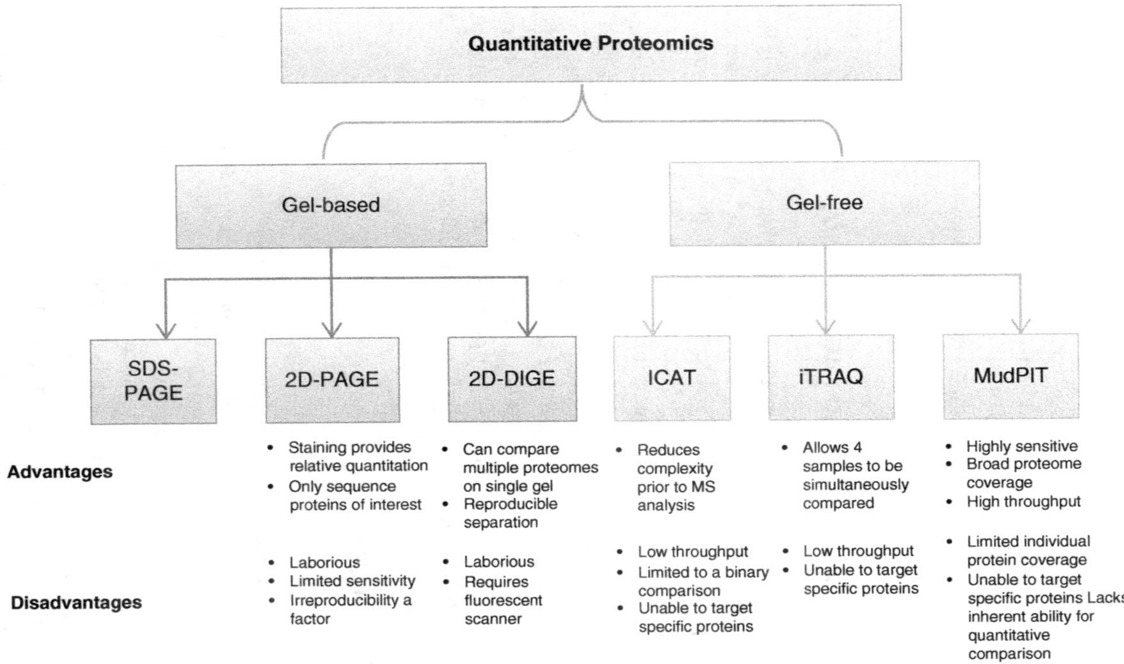

Figure 7.3 Most commonly used gel-based and gel-free quantitative proteomics with their specific advantages and disadvantages.

associated with active metabolic changes (Tsunezuka et al. 2005). The iTRAQ approach was recently used to identify the proteins and pathways associated with the lesion mimic mutant E40 in rice, and it was found that 233 proteins involved in diverse biological processes and phenylpropanoid biosynthesis were significantly up- or down-regulated in E40 compared with wild type (Yang et al. 2019). Proteomic investigation focusing on the defense signaling pathway in plants was also carried out by other scientists (Kim et al. 2003; Peck et al. 2001) (Figure 7.3).

7.7 Conclusion and Future Perspective

It can be concluded from this chapter that proteomics is an important alterative and advanced approach to genomics which can provide better insight into plant biology. Recent advancement in proteomics has helped us to understand plant biology clearly. But there is still more information about these advanced techniques needed to overcome their limitations to have a clear scenario of plant response mechanisms to stress conditions. The PTMs have made it necessary to study proteins at each stage. Thus, having accurate and complete knowledge of the proteins involved in plant biology is important to develop smart crops which can withstand multiple stresses at the same time.

References

Abdallah, C., Dumas-Gaudot, E., Renaut, J., and Sergeant, K. (2012). Gel-based and gel-free quantitative proteomics approaches at a glance. *International Journal of Plant Genomics* https://doi.org/10.1155/2012/494572.

Afzal, A.J., Natarajan, A., Saini, N. et al. (2009). The nematode resistance allele at the *rhg1* locus alters the proteome and primary metabolism of soya bean roots. *Plant Physiology* 151 (3): 1264–1280. https://doi.org/10.1104/pp.109.138149.

Agarwal, K., Choe, L.H., and Lee, K.H. (2006). Shotgun proteomics using the iTRAQ isobaric tags. *Briefings in Functional Genomics & Proteomics* 5 (2): 112–120. https://doi.org/10.1093/bfgp/ell018.

Alvarez, S., Zhu, M., and Chen, S. (2009). Proteomics of *Arabidopsis* redox proteins in response to methyl jasmonate. *Journal of Proteomics* 73: 30–40.

Babujee, L., Venkatesh, B., Yamazaki, A., and Tsuyumu, S. (2007). Proteomic analysis of the carbonate insoluble outer membrane fraction of the soft-rot pathogen *Dickeya dadantii* (syn. *Erwinia chrysanthemi*) strain 3937. *Journal of Proteome Research* 6 (1): 62–69. https://doi.org/10.1021/pr060423l.

Bohler, S., Sergeant, K., Hoffmann, L. et al. (2011). A difference gel electrophoresis study on thylakoids isolated from poplar leaves reveals a negative impact of ozone exposure on membrane proteins. *Journal of Proteome Research* 10 (7): 3003–3011.

Buensanteai, N., Yuen, G.Y., and Prathuangwong, S. (2009). Priming, signaling and protein production associated with induced resistance by *Bacillus amyloliquefaciens* KPS46. *World Journal of Microbiology and Biotechnology* 25 (7): 1275–1286. https://doi.org/10.1007/s11274-009-0014-6.

Campo, S., Carrascal, M., Coca, M. et al. (2004). The defence response of germinating maize embryos against fungal infection: a proteomics approach. *Proteomics* 4 (2): 383–396. https://doi.org/10.1002/pmic.200300657.

Campos, A., da Costa, G., Coelho, A.V., and Fevereiro, P. (2009). Identification of bacterial protein markers and enolase as a plant response protein in the infection of *Olea europaea* subsp. *europaea* by *Pseudomonas savastanoi* pv. *savastanoi*. *European Journal of Plant Pathology* 125 (4): 603–616. https://doi.org/10.1007/s10658-009-9509-0.

Castillejo, M.A., Maldonado, A.M., Dumas-Gaudot, E. et al. (2009). Differential protein expression to investigate responses and resistance to *Orobranche crenata* in *Medicago truncatula*. *BMC Genomics* 10: 294–298. https://doi.org/10.1186/1471-2164-10-294.

Cecconi, D., Orzetti, S., Vandelle, E. et al. (2009). Protein nitration during defense response in *Arabidopsis thaliana*. *Electrophoresis* 30 (14): 2460–2468. https://doi.org/10.1002/elps.200800826.

Chivasa, S., Hamilton, J.M., Pringle, R.S. et al. (2006). Proteomic analysis of differentially expressed proteins in fungal elicitor-treated *Arabidopsis* cell cultures. *Journal of Experimental Botany* 57 (7): 1553–1562. https://doi.org/10.1093/jxb/erj149.

Coaker, G.L., Willard, B., Kinter, M. et al. (2004). Proteomic analysis of resistance mediated by Rcm 2.0 and Rcm 5.1, two loci controlling resistance to bacterial canker of tomato. *Molecular Plant-Microbe Interactions* 17 (9): 1019–1028. https://doi.org/10.1094/MPMI.2004.17.9.1019.

Colditz, F., Nyamsuren, O., Niehaus, K. et al. (2004). Proteomic approach: identification of *Medicago truncatula* proteins induced in roots after infection with the pathogenic oomycete *Aphanomyces euteiches*. *Plant Molecular Biology* 55 (1): 109–120. https://doi.org/10.1007/s11103-004-0499-1.

Collins, R.M., Afzal, M., Ward, D.A. et al. (2010). Differential proteomic analysis of *Arabidopsis thaliana* genotypes exhibiting resistance or susceptibility to the insect herbivore, *Plutella xylostella*. *PLoS One* 5: e10103.

Dahal, D., Heintz, D., Van Dorsselaer, A. et al. (2009). Pathogenesis and stress related, as well as metabolic proteins are regulated in tomato stems infected with *Ralstonia solanacearum*. *Plant Physiology and Biochemistry* 47 (9): 838–846. https://doi.org/10.1016/j.plaphy.2009.05.001.

De Wit, P.J. (2007). How plants recognize pathogens and defend themselves. *Cellular and Molecular Life Sciences* 64 (21): 2726–2732. https://doi.org/10.1007/s00018-007-7284-7.

Delalande, F., Carapito, C., Brizard, J.P. et al. (2005). Multigenic families and proteomics: extended protein characterization as a tool for paralog gene identification. *Proteomics* 5 (2): 450–460. https://doi.org/10.1002/pmic.200400954.

Diaz-Vivancos, P., Rubio, M., Mesonero, V. et al. (2006). The apoplastic antioxidant system in *Prunus*: response to long-term plum pox virus infection. *Journal of Experimental Botany* 57 (14): 3813–3824. https://doi.org/10.1093/jxb/erl138.

Dodds, P.N. and Rathjen, J.P. (2010). Plant immunity: towards an integrated view of plant–pathogen interactions. *Nature Reviews Genetics* 11 (8): 539–548. https://doi.org/10.1038/nrg2812.

Duceppe, M.O., Cloutier, C., and Michaud, D. (2012). Wounding, insect chewing and phloem sap feeding differentially alter the leaf proteome of potato, *Solanum tuberosum* L. *Proteome Science* 10: 73.

Duley, H. and Grover, A. (2001). Current initiatives in proteomics research: the plant perspective. *Current Science* 80 (2): 262–269.

Durand, T.C., Sergeant, K., Planchon, S. et al. (2010). Acute metal stress in *Populus tremula* x *P. alba* (717-1B4 genotype): leaf and cambial proteome changes induced by cadmium^{2+}. *Proteomics* 10 (3): 349–368.

Eubel, H., Braun, H.P., and Millar, A.H. (2005). Blue-native PAGE in plants: a tool in analysis of protein–protein interactions. *Plant Methods* 1 (1): 11.

Franco, O.L., Pereira, J.L., Costa, P.H.A. et al. (2010). Methodological evaluation of 2DE to study root proteomics during nematode infection in cotton and coffee plants. *Preparative Biochemistry and Biotechnology* 40 (2): 152–163. https://doi.org/10.1080/10826060903558976.

Gordon, J.A. and Jencks, W.P. (1963). The relationship of structure to the effectiveness of denaturing agents for proteins. *Biochemistry* 2 (1): 47–57.

Gorg, A., Weiss, W., and Dunn, M.J. (2004). Current two-dimensional electrophoresis: technology for proteomics. *Proteomics* 4 (12): 3665–3685. https://doi.org/10.1002/pmic.200401031.

Gygi, S.P., Rist, B., Gerber, S.A. et al. (1999). Quantitative analysis of complex protein mixtures using isotope-coded affinity tags. *Nature Biotechnology* 17 (10): 994–999. https://doi.org/10.1038/13690.

Gygi, S.P., Corthals, G.L., Zhang, Y. et al. (2000). Evaluation of two-dimensional gel electrophoresis based proteome analysis technology. *Proceedings of the National Academy of Sciences of the United States of America* 97 (17): 9390–9395.

Houterman, P.M., Speijer, D., Dekker, H.L. et al. (2007). The mixed xylem sap proteome of *Fusarium oxysporum*-infected tomato plants. *Molecular Plant Pathology* 8 (2): 215–221. https://doi.org/10.1111/j.1364-3703.2007.00384.x.

Jones, A.M.E., Thomas, V., Bennett, M.H. et al. (2006). Modifications to the *Arabidopsis* defense proteome occur prior to significant transcriptional change in response to inoculation with *Pseudomonas syringae*. *Plant Physiology* 142 (4): 1603–1620. https://doi.org/10.1104/pp.106.086231.

Kieffer, P., Dommes, J., Hoffmann, L. et al. (2008). Quantitative changes in protein expression of cadmium-exposed poplar plants. *Proteomics* 8 (12): 2514–2530.

Kim, S., Cho, K., Yu, S. et al. (2003). Proteomic analysis of differentially expressed proteins induced by rice blast fungus and elicitor in suspension-cultured rice cells. *Proteomics* 3: 2368–2378.

Kim, S.T., Kim, S.G., Hwang, D.H. et al. (2004). Proteomic analysis of pathogen-responsive proteins from rice leaves induced by rice blast fungus, *Magnaporthe grisea*. *Proteomics* 4 (11): 3569–3578. https://doi.org/10.1002/pmic.200400999.

Klose, J. (1975). Protein mapping by combined isoelectric focusing and electrophoresis of mouse tissues: a novel approach to testing for induced point mutation in mammals. *Human Genetics* 26 (3): 231–243.

Kundu, S. et al. (2011). Proteomic analysis of salicylic acid induced resistance to mung bean yellow mosaic India virus in *Vigna mungo*. *Journal of Proteomics* 74 (3): 337–349. https://doi.org/10.1016/j.jprot.2010.11.012.

Lambert, J.P., Ethier, M., Smith, J.C., and Figeys, D. (2005). Proteomics from gel based to gel free. *Analytical Chemistry* 77 (12): 3771–3787. https://doi.org/10.1021/ac050586d.

Lee, J., Feng, J., Campbell, K.B. et al. (2009). Quantitative proteomic analysis of bean plants infected by a virulent and avirulent obligate rust fungus. *Molecular & Cellular Proteomics* 8: 19–31. https://doi.org/10.1074/mcp.M800156-MCP200.

Li, T., Xu, S.L., Oses-Prieto, J.A. et al. (2011). Proteomics analysis reveals post-translational mechanisms for cold-induced metabolic changes in *Arabidopsis*. *Molecular Plant* 4 (2): 361–374.

Liang, Y., Srivastava, S., Rahman, M.H. et al. (2008). Proteome changes in leaves of *Brassica napus* L. as a result of *Sclerotinia sclerotiorum* challenge. *Journal of Agricultural and Food Chemistry* 56 (60): 1963–1976. https://doi.org/10.1021/jf073012d.

Lilley, K.S., Razzaq, A., and Dupree, P. (2002). Two-dimensional gel electrophoresis: recent advances in sample preparation, detection and quantitation. *Current Opinion in Chemical Biology* 6 (1): 46–50.

Lodha, T.D. and Basak, J. (2012). Plant–pathogen interaction: what microarray tells about it? *Molecular Biotechnoogy* 50 (1): 87–97.

Lodha, T.D., Hembram, P., Tep, N., and Basak, J. (2013). Proteomics: a successful approach to understand the molecular mechanism of plant–pathogen interaction. *American Journal of Plant Sciences* 4: 1212–1226. https://doi.org/10.4236/ajps.2013.46149.

Lund, T.C., Anderson, L.B., McCullar, V. et al. (2007). iTRAQ is a useful method to screen for membrane bound proteins differentially expressed in human natural killer cell types. *Journal of Proteome Research* 6 (2): 644–653. https://doi.org/10.1021/pr0603912.

Mahmood, T., Jan, A., Kakishima, M., and Komatsu, S. (2006). Proteomic analysis of bacterial-blight defence responsive proteins in rice leaf blades. *Proteomics* 6 (22): 6053–6065. https://doi.org/10.1002/pmic.200600470.

Marouga, R., David, S., and Hawkins, E. (2005). The development of the DIGE system: 2D fluorescence difference gel analysis technology. *Analytical and Bioanalytical Chemistry* 382 (3): 669–678. https://doi.org/10.1007/s00216-005-3126-3.

Mehta, A. and Rosato, Y.B. (2001). Differentially expressed proteins in the interaction of *Xanthomonas axonopodis* pv. *citri* with leaf extract of the host plant. *Proteomics* 1 (9): 1111–1118.

Meng, Q., Rao, L., Xiang, X. et al. (2011). A systematic strategy for proteomic analysis of chloroplast protein complexes in wheat. *Bioscience, Biotechnology, and Biochemistry* 75 (11): 2194–2199.

Mukherjee, A.K., Carp, M.J., Zuchman, R. et al. (2010). Proteomics of the response of *Arabidopsis thaliana* to infection with *Alternaria brassicicola*. *Journal of Proteomics* 73 (4): 709–720. https://doi.org/10.1016/j.jprot.2009.10.005.

Nat, N.V.K., Srivastava, S., Yajima, W., and Sharma, N. (2007). Application of proteomics to investigate plant–pathogen interactions. *Current Proteomics* 4 (1): 28–43. https://doi.org/10.2174/157016407781387357.

O'Farrell, P.H. (1975). High resolution two dimensional electrophoresis of proteins. *Journal of Biological Chemistry* 250 (10): 4007–4021.

Ong, S.E. and Pandey, A. (2001). An evaluation of the use of two dimensional gel electrophoresis in proteomics. *Biomolecular Engineering* 18 (5): 195–205.

Park, O.K. (2004). Proteomic study in plants. *Journal of Biochemistry and Molecular Biology* 37 (1): 133–138. https://doi.org/10.5483/BMBRep.2004.37.1.133.

Patterson, S.D. and Aebersold, R.H. (2003). Proteomics: the first decade and beyond. *Nature Genetics* 33 (3): 311–321. https://doi.org/10.1038/ng1106.

Peck, S.C., Nuhse, T.S., Hess, D. et al. (2001). Directed proteomics identifies a plant-specific protein rapidly phosphorylated in response to bacterial and fungal elicitors. *Plant Cell* 13: 1467–1475.

Perez-Bueno, M.L., Rahoutei, J., Sajnani, C. et al. (2004). Proteomic analysis of the oxygen-evolving complex of photosystem II under biotic stress: studies on *Nicotiana benthamiana* infected with tobamoviruses. *Proteomics* 4 (20): 418–425. https://doi.org/10.1002/pmic.200300655.

Rampitsch, C., Bykova, N.V., Mccallum, B. et al. (2006). Analysis of the wheat and *Puccinia triticina* (leaf rust) proteomes during a susceptible host pathogen interaction. *Proteomics* 6 (6): 1897–1907. https://doi.org/10.1002/pmic.200500351.

Righetti, P.G., Castagna, A., Antonucci, F. et al. (2004). Critical survey of quantitative proteomics in two-dimensional electrophoretic approaches. *Journal of Chromatography A* 1051 (1): 3–17. https://doi.org/10.1016/j.chroma.2004.05.106.

Rose, J.K.C., Bashir, S., Giovannoni, J.J. et al. (2004). Tackling the plant proteome: practical approaches, hurdles and experimental tools. *Plant Journal* 39 (5): 715–733. https://doi.org/10.1111/j.1365-313X.2004.02182.x.

Ross, P.L., Huang, Y.N., Marchese, J.N. et al. (2004). Multiplexed protein quantitation in *Saccharomyces cerevisiae* using amine-reactive isobaric tagging reagents. *Molecular & Cellular Proteomics* 3 (12): 1154–1169. https://doi.org/10.1074/mcp.M400129-MCP200.

Schagger, H. and Von Jagow, G. (1991). Blue native electrophoresis for isolation of membrane protein complexes in enzymatically active form. *Analytical Biochemistry* 199 (2): 223–231.

Schenkluhn, L., Hohnjec, N., Niehaus, K. et al. (2010). Differential gel electrophoresis (DIGE) to quantitatively monitor early symbiosis and pathogenesis-induced changes of the *Medicago truncatula* root proteome. *Journal of Proteomics* 73 (4): 753–768.

Schulze, W.X. and Usadel, B. (2010). Quantitation in mass spectrometry based proteomics. *Annual Review of Plant Biology* 61: 491–516.

Segarra, G., Casanova, E., Bellido, D. et al. (2007). Proteome, salicylic acid and Jasmonic acid changes in cucumber plants inoculated with *Trichoderma asperellum* strain T34. *Proteomics* 7 (21): 3942–3952. https://doi.org/10.1002/pmic.200700173.

Sergeant, K., Spieß, N., Renaut, J. et al. (2011). One dry summer: a leaf proteome study on the response of oak to drought exposure. *Journal of Proteomics* 74 (8): 1385–1395.

Sharma, N., Hotte, N., Rahman, M.H. et al. (2008). Towards identifying *Brassica* proteins involved in mediating resistance to *Leptosphaeria maculans*: a proteomics-based approach. *Proteomics* 8 (17): 3516–3535. https://doi.org/10.1002/pmic.200701141.

Tahara, S.T., Mehta, A., and Rosato, Y.B. (2003). Proteins induced by *Xanthomonas axonopodis* pv. *passiflorae* with leaf extract of the host plant (*Passiflorae edulis*). *Proteomics* 3 (1): 95–102.

Thelen, J.J. and Peck, S.C. (2007). Quantitative proteomics in plants: choices in abundance. *Plant Cell* 19 (11): 3339–3346.

Thurston, G., Regan, S., Rampitsch, C., and Xing, T. (2005). Proteomic and phosphoproteomic approaches to understand plant-pathogen interactions. *Physiological and Molecular Plant Pathology* 66 (1): 3–11. https://doi.org/10.1016/j.pmpp.2005.03.004.

Tonge, R.P., Shaw, J., Middleton, B. et al. (2001). Validation and development of fluorescence two-dimensional differential gel electrophoresis proteomics technology. *Proteomics* 1 (3): 377–396.

Tsunezuka, H., Fujiwara, M., Kawasaki, T., and Shimamoto, K. (2005). Proteome analysis of programmed cell death and defense signaling using the rice lesion mimic mutant *cdr2*. *The American Phytopathological Society* 18 (1): 52–59.

Unlu, M., Morgan, M.E., and Minden, J.S. (1997). Difference gel electrophoresis: a single gel method for detecting changes in protein extracts. *Electrophoresis* 18 (11): 2071–2077. https://doi.org/10.1002/elps.1150181133.

Valcu, C.M., Junqueira, M., Shevchenko, A., and Schlink, K. (2009). Comparative proteomic analysis of responses to pathogen infection and wounding in *Fagus sylvatica*. *Journal of Proteome Research* 8 (8): 4077–4091. https://doi.org/10.1021/pr900456c.

Van den Bergh, G. and Arckens, L. (2005). Recent advances in 2D electrophoresis: an array of possibilities. *Expert Review of*

Proteomics 2 (2): 243–252. https://doi.org/10.1586/14789450.2.2.243.

Van Noorden, G.E., Kerim, T., Goffard, N. et al. (2007). Overlap of proteome changes in *Medicago truncatula* in response to auxin and *Sinorhizobium meliloti*. *Plant Physiology* 144 (2): 1115–1131.

Vanderschuren, H., Lentz, E., Zainuddin, I., and Gruissem, W. (2013). Proteomics of model and crop plant species: status, current limitations and strategic advances for crop improvement. *Journal of Proteomics* 93: 5–19.

Ventelon-Debout, M., Delalande, F., Brizard, J.P. et al. (2004). Proteome analysis of cultivar-specific deregulations of *Oryza sativa indica* and *O. sativa japonica* cellular suspension undergoing *rice yellow mottle virus* infection. *Proteomics* 4 (1): 216–225. https://doi.org/10.1002/pmic.200300502.

Vertommen, A., Panis, B., Swennen, R., and Carpentier, S.C. (2011). Challenges and solutions for the identification of membrane proteins in non-model plants. *Journal of Proteomics* 74 (8): 1165–1181.

Wei, Z., Hu, W., Lin, Q. et al. (2009). Understanding rice plant resistance to the brown planthopper (*Nilaparvata lugens*): a proteomic approach. *Proteomics* 9 (10): 2798–2808. https://doi.org/10.1002/pmic.200800840.

Wilkins, M.R., Scnchez, J.C., Gooley, A.A. et al. (1996). Progress with proteome projects: why all proteins expressed by a genome should be identified and how to do it. *Biotechnology & Genetic Engineering Reviews* 13 (1): 19–50. https://doi.org/10.1080/02648725.1996.10647923.

Wittmann-Liebold, B., Graack, H.R., and Pohl, T. (2006). Two-dimensional gel electrophoresis as tool for proteomics studies in combination with protein identification by mass spectrometry. *Proteomics* 6 (17): 4688–4703. https://doi.org/10.1002/pmic.200500874.

Wongpia, A. and Lomthaisong, K. (2010). Changes in the 2DE protein profiles of chili pepper. *Science Asia* 36: 259–270.

Yang, X.-B., Meng, W.-L., Zhao, M.-J. et al. (2019). Pathways associated with the novel lesion mimic mutant E40 in rice using iTRAQ-based strategy. *International Journal of Molecular Sciences* 20: 1294. https://doi.org/10.3390/ijms20061294.

Zhang, J.H., Sun, L.W., Liu, L.L. et al. (2010). Proteomic analysis of interactions between the generalist herbivore *Spodoptera exigua* (Lepidoptera: Noctuidae) and *Arabidopsis thaliana*. *Plant Molecular Biology Reporter* 28 (2): 324–333. https://doi.org/10.1007/s11105-009-0156-6.

Zhou, W., Eudes, F., and Laroche, A. (2006). Identification of differentially regulated proteins in response to a compatible interaction between the pathogen *Fusarium graminearum* and its host, *Triticum aestivum*. *Proteomics* 6 (16): 4599–4609. https://doi.org/10.1002/pmic.200600052.

Zhu, H., Bilgin, M., and Snyder, M. (2003). Proteomics. *Annual Review of Biochemistry* 72 (1): 783–812. https://doi.org/10.1146/annurev.biochem.72.121801.161511.

Zieske, L.R. (2006). A perspective on the use of iTRAQ reagent technology for protein complex and profiling studies. *Journal of Experimental Botany* 57 (7): 1501–1508. https://doi.org/10.1093/jxb/erj168.

8

RNA World and Heat Stress Tolerance in Plants

Usman Ijaz[1], *Muhammad Amjad Ali*[2], *Habibullah Nadeem*[1], *Lin Tan*[3], *and Farrukh Azeem*[1,3]

[1] *Department of Bioinformatics and Biotechnology, Government College University Faisalabad (GCUF), Faisalabad, Pakistan*
[2] *Department of Plant Pathology, University of Agriculture, Faisalabad, Pakistan*
[3] *Haikou Experimental Station, Chinese Academy of Tropical Agricultural Sciences (CATAS), Hainan Key Laboratory of Banana Genetic Improvement, Haikou, Hainan Province, China*

8.1 Introduction

Environmental constraints like drought, salinity, and temperature stress can severely affect plant growth and agronomic yields. However, plants have developed highly specialized mechanisms to survive unfavorable conditions (Banerjee et al. 2017; Rai et al. 2018). Since climate change is promoting unforeseen temperature variations, the scientific community has increased interest in understanding plant responses to temperature stress. Heat stress negatively affects cellular homeostasis and photosynthetic activity, promotes growth retardation, and may result in plant death (Liu et al. 2014). A number of studies have highlighted the regulatory mechanisms, which involve plant response to heat stress through genes or transcription factors (Ohama et al. 2017). It may include heat stress transcription factors (HSFs) and heat-responsive genes (Huang and Xu 2008). With the availability of sophisticated and readily available sequencing technologies, a number of recent studies have elaborated the importance of noncoding RNAs (ncRNAs) in plants' response to high temperature.

Heat Stress Tolerance in Plants: Physiological, Molecular and Genetic Perspectives, First Edition.
Edited by Shabir Hussain Wani and Vinay Kumar.

The ncRNAs are unable to encode functional proteins and include miRNAs (microRNAs), siRNAs (small interfering RNAs), lncRNAs (long noncoding RNAs), and circRNAs (circular RNAs) (Borges and Martienssen 2015). The ncRNAs are considered critical components of cellular machinery that regulate gene expression before and after transcription, stability of RNA, and even translation (Khraiwesh et al. 2012; de Lima et al. 2012). As several studies have reviewed biogenesis and characteristic features of ncRNAs (Zhang 2015). This chapter highlights the importance of miRNAs, siRNAs, lncRNAs, and circular RNAs for heat stress response of land plants (Table 8.1).

8.2 Plant microRNAs

Plant miRNAs are generally 21–24 nucleotides (nt) in length and negatively regulate the expression of a gene during development, metabolism, and plant response to unfavorable environmental conditions (Bulgakov and Avramenko 2015; Zhao et al. 2016). Similarly, plant parasitic nematodes manipulate various miRNA pathways to develop successful parasitism in plants (Ali et al. 2017; Ali et al. 2018). Unlike other types of ncRNAs, the formation of mature miRNA is catalyzed by a complex enzymatic process, from the initial transcript pre-miRNA to the shorter, stem loop RNAs or precursor (pre-miRNA) with a characteristic hairpin structure, followed by miRNA duplex (miRNA: miRNA*). Numerous key enzymes, e.g. Dicer-like 1 (DCL1) enzyme, are involved in the maturation of miRNAs in plants (Liu et al. 2018; Zhao et al. 2016). These ncRNAs regulate the expression of genes in multiple ways, including degrading mRNA or inhibiting the translation process (Prasad and Jagadish 2015). Although, several miRNAs that respond to heat stress in different plant species have been reported.

miR398 is a well-studied miRNA in plant species, which responds to multiple stresses (Noman et al. 2017). In *Arabidopsis*, miR398 regulates a copper chaperone for SOD (*CCS1*), a subunit of the mitochondrial cytochrome *c* oxidase (*Cox5b-1*) and closely related copper/zinc superoxide dismutases (*CSD1* and *CSD2*) (Zhu et al. 2011; Borges and Martienssen 2015).

Table 8.1 Summary of ncRNAs expressed under heat stress in different plant species.

Serial number	miRNA	Target	Plant species	Reference
1.	miR156	SPL	*Arabidopsis, Zea mays, Populus trichocarpa, Musa acuminata*	Kim et al. (2012) Xin et al. (2010); Yu et al. (2012); Stief et al. (2014); Zhu et al. (2019)
2.	miR159	MYB	*Phaseolus vulgaris, Arabidopsis*	Hivrale et al. (2016); Ohama et al. (2017); Liu et al. (2017)
3.	miR160	ARF	*Manihot esculenta, Populus euphratica*	Xin et al. (2010); Chen et al. (2012); May et al. (2013); Hivrale et al. (2016); Lin et al. (2018)
4.	miR164	NAC	*Vigna unguiculata*	M. Y. Li et al. (2014); Kruszka et al. (2014); Hivrale et al. (2016)
5.	miR165	HD-ZIPIII	*Arabidopsis thaliana, Hordeum vulgare, Triticum turgidum*	Chen et al. (2012); Guan et al. (2013); Smoczynska et al. (2019)
6.	miR167	ARF	*Oryza sativa*	Lu et al. (2008); Chen et al. (2012); Mahale et al. (2013)
7.	miR168	**AGO-1**	*Arabidopsis thaliana, Oryza sativa, Populus euphratica*	Sailaja et al. (2013); Hivrale et al. (2016); M. Y. Li et al. (2014)

(Continued)

Table 8.1 (Continued)

Serial number	miRNA	Target	Plant species	Reference
8.	miR169	NF-Y	*Arabidopsis thaliana*	Hivrale et al. (2016); May et al. (2013); Khaksefidi et al. (2015)
9.	miR170/ miR171	**SCL**	*Populus trichocarpa*	Kruszka et al. (2014); Yao et al. (2010); Khaksefidi et al. (2015)
10.	miR172	AP2	*Arabidopsis*	Mishra & Bohra (2018); Lu et al. (2008); Chen et al. (2012)
11.	miR390	ARF	*Populus euphratica*	Kumar et al. (2014); Hivrale et al. (2016)
12.	miR319	TCP	*Arabidopsis, Oryza sativa, Populus euphratica*	Schommer et al. (2008) Giacomelli et al. (2012); Smoczynska et al. (2019)
13.	miR393	TIRI/ARF	*Triticum turgidum, Oryza sativa, Agrostis*	Shriram et al. (2016); Mahale et al. (2013); Zhao et al. (2019)
14.	miR395	F-box	*Glycine max, T. aestivum*	Xin et al. (2010); Chen et al. (2012); Hivrale et al. (2016); Banerjee et al. (2017)
15.	miR396	GRF/bHLH /WRKY	*Oryza sativa, Glycine max, Triticum aestivum*	S. X. Li et al. (2014); Megha et al. (2018)

16.	miR397	Laccases	*Arabidopsis, Oryza sativa*	Sailaja et al. (2013); Khaksefidi et al. (2015); Hivrale et al. (2016)
17.	miR398	CSD, CCS, COX5	*Arabidopsis, Oryza sativa, Triticum aestivum*	Hivrale et al. (2016); Kumar (2018)
18.	miR400	PPR	*Arabidopsis thaliana*	S. X. Li et al. (2014); Yan et al. (2012)
19.	MiR5175	ACC like oxidase	*Hordeum vulgare*	Kruszka et al. (2014); Hivrale et al. (2016)
20.	siRNA 005047_0654_1954.1		*Triticum aestivum*	Xin et al. (2010); Yu et al. (2012)
21.	SiRNA 080621_1340_98.1		*Triticum aestivum*	Yao et al. (2010)
22.	TAS1-siRNAs	HTT1, HTT2	*Arabidopsis thaliana*	S. X. Li et al. (2014)
23.	lnRNA5		*Triticum aestivum*	Xin et al. (2011)
24.	lnRNA27		*Triticum aestivum*	Xin et al. (2011)
25.	circRNA 2:13427439-13429199		*Arabidopsis thaliana*	Pan et al. (2018)
26.	circRNA 3:4390903-4391594		*Arabidopsis thaliana*	Pan et al. (2018)

NF-y, nuclear transcription factor Y; SBP, squamosa binding promoter; APS/AST, ATP sulfurylase/affinity sulfate transporter; TIR1, transport inhibitor response 1, AFB, auxin F-box; bHLH, basic helix loop helix; GRF, growth hormone releasing factor.

Among these, CSDs act as reactive oxygen species (ROS) scavengers, while *CSD/CCS* negatively regulates ROS species, which are also related with HSP and HSF synthesis (Guan et al. 2013; Lu et al. 2013). It was observed that miR398 expression is upregulated under elevated temperature stress while target genes (*CCS, CSD1,* and *CSD2*) are downregulated. Transgenic plants expressing mutated *CSD1, CSD2,* or *CCS* (miR398-resistant versions) were hypersensitive to heat stress. Alternatively, *csd1, csd2,* or *ccs* plants showed increased HSF/HSP levels and heat-stress tolerance (Lu et al. 2013; Guan et al. 2013; Kumar 2018). Similar results were obtained for miR398-mediated regulation of *CDSs* in *Populus tomentosa* and *Brassica rapa,* which indicate conserved role of miR398 in thermos-tolerance of plants ((Kotak et al. 2007; Yu et al. 2012; Ohama et al. 2017).

miR156 and miR172 negatively affect flowering and vegetative phase change by targeting *SPL* (*squamosa promoter binding protein-like*) and *AP2* (*APETALA2)* genes (Wu et al. 2009). When plants face heat stress, miR156 expression is significantly up-regulated and compared to a down-regulation of *SPL*. It further stimulates the expression of *FLU* (*FRUITULL*) and *FT* (*FLOWERING LOCUS T*) in *Arabidopsis* (Figure 8.1). When *B. rapa* faces heat stress, miR156g and miR156h expression is increased and its target gene (*BracSPL2*) is decreased (Zhu et al. 2019). Interestingly, in *Arabidopsis,* an isoform of miR156 is essential for heat stress memory. Similarly, miR156 can specifically promote *SPL2/SPL11*-mediated continuous expression of heat responsive genes (Cui et al. 2014; Stief et al. 2014). Therefore, it is anticipated that highly conserved miR156 (and its target genes) is important for heat stress memory in plants (Stief et al. 2014; Ohama et al. 2017).

miR172 can target *AP2* like genes including *SMZ (SCHLAFMUTZE)* and *TOE1/TOE2* (*TARGET OF EAT1*) (Chen et al. 2012; Mishra and Bohra 2018). In sunflower (*Helianthus annuus), Arabidopsis,* and wheat plants, miR172 is downregulated in heat stress as compared to miR156 (Liu et al. 2017). Similarly, high temperature suppresses the expression of miR172 and up-regulates the expression level of its target gene *TOE2* (M. Y. Li et al. 2014; S. X. Li et al. 2014). Eventually, these results show that elevated temperature can modulate the expression *miR156–SPL–miR172–AP2* pathway genes (May et al. 2013).

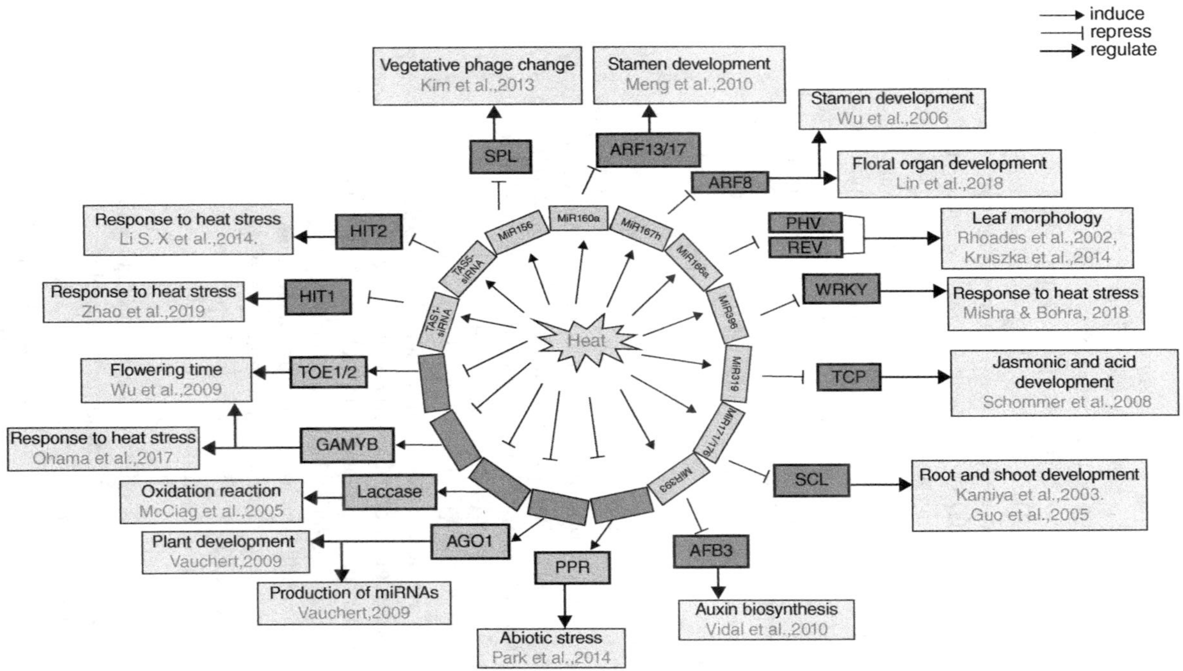

Figure 8.1 An overview of noncoding RNAs regulated under heat stress in plants. A miRNA-target network module involved in the heat stress response based on the changes in expression profiles of miRNAs and their targets in plants under heat stress. Black arrows represent regulatory effects (position or negative regulation); green boxes: upregulated; red boxes: downregulated. Source: Adapted from Zhao et al. (2016).

Phytohormones (abscisic acid, auxins, ethylene, and jasmonate) are essential for plant responses to environmental stresses. Auxin is involved in plant growth by monitoring the elongation and division of cells (de Wit et al. 2014; Gallavotti 2013). Since heat stress primarily hinders plant growth, the auxin signaling pathway is considered important for plant thermotolerance. miR160 (which is involved in auxin signaling) regulates auxin response factor (*ARF*) genes, that participate in development of roots, shoots, and flowers (Meng et al. 2010; Lin et al. 2018). It was noticed that miR160 was up-regulated and *ARFs* were downregulated in response to high temperature (Khaksefidi et al. 2015). Though a contradictory regulation was observed in wheat, when miR160 was down-regulated compared to *HSP70* (target protein) under heat stress (Kumar et al. 2015). Similarly, miR167 regulates *ARF8* (involved in development of floral organs), but its expression is decreased in response to elevated temperature (Kruszka et al. 2014). Additionally, there are some other miRNAs (miR390 and miR393) that are involved in auxin signaling pathway and heat stress response (Hivrale et al. 2016). The *GAMYB* (gibberellic acid MYB) genes play a key role in germination and the development of flowers and are negatively regulated by miR159 (Reyes and Chua 2007). It has been reported that under heat stress conditions the expression of miR159 was decreased, while *TaGAMYB* was up-regulated (Wang et al. 2012). A series of experiments involving rice and *Arabidopsis* demonstrated that miR159 overexpression promoted down-regulation of *GAMYB* and hence decreased plant thermotolerance. Moreover, TCPs (targets of miR319) regulate jasmonic acid biosynthesis. It has been reported that under heat stress conditions miR319 was up-regulated and *TCPs* (teosinte branched/cycloidea proteins) were down-regulated (Hivrale et al. 2016). Such reports are important indications for hormones regarding the miRNA-mediated heat stress responses of plants.

The most important effect of heat stress is to hinder photosynthesis, seed germination, and the development of plants. Several microRNAs have been reported which play significant role in plant development. The miR164-regulated NAC TFs (nascent polypeptide associated complex) perform essential functions for root and shoot development (Guo et al. 2005).

In response to heat stress, *AtNAC1* expression is suppressed when miR164 is inducted (M. Y. Li et al. 2014; S. X. Li et al. 2014). miR164 can specifically activate the heat-shock protein (*HSP17*) under heat stress conditions. It is observed that *HSP17* is up-regulated in heat-stressed wheat plants (Kumar et al. 2015). miR166 can target HD-Zip (homeodomain leucine zipper) transcription factors, which regulate leaf morphology and initiation of auxiliary meristem (Rhoades et al. 2002). In *Hordeum vulgare*, miR166a expression is highly induced, while its target genes *HOX9* (homeobox leucine-zipper protein), *REV* (REVOLUTA), and *PHV* (PHAVOLUTA) are down-regulated in heat stress response (Kruszka et al. 2014).

Similarly, miR171 is crucial for the expression of GRAS family genes (like *SCL6-IV*) and *SCL6-III* (SCARECROW-LIKE6-III), which play a role in plant development like radial patterning of shoots and roots (Kamiya et al. 2003; Khaksefidi et al. 2015). In *Arabidopsis*, miR171 is up-regulated, while the expression of GRAS genes is suppressed in response to heat stress (Barku et al. 2013). Although, contrary results are reported in *Populus*, where Ptc-miR171 and Pto-171 are down-regulated under heat stress conditions (Chen et al. 2012; Lu et al. 2008). miR168 is another MiRNA which is important for plant development. It targets *AGO1* (*ARGONATUE*) (Vaucheret 2009). However, miR168 is down-regulated under heat stress, which leads to an up-regulation of *AGO1* (Chen et al. 2012). Likewise, there are additional examples of miRNAs-target genes – miR396-*HaWRKY6*, miR397-*LACCASE*, miR400-*PPR* (*pentatricopeptide repeat*) – which play a role in plant development and heat stress response (Giacomelli et al. 2012; Yan et al. 2012).

Alternative splicing (AS) is one of the key mechanisms which add diversity to transcriptomes and proteomes of plants. Previous studies have reported that several genes with higher AS can play an important role in plant response to environmental stresses (Guerra et al. 2015; Shang et al. 2017). In *Zea mays*, the expression of *DREB2B* (dehydration responsive element binding 2B) is regulated AS under heat-stressed conditions (Qin et al. 2007; Matsukura et al. 2010). Likewise, AS also regulates the expression of miRNAs to protect plants from heat stress. Intronic miR400 is transcribed along the host gene and its expression is reduced under heat stress while primary transcript of miR400 is up-regulated. In addition, *Arabidopsis* seeds with overexpressing miR400 exhibit a low germination rate under heat stress (Yan et al. 2012). Interestingly, *PPR* (a target of miR400) is also up-regulated under heat stress in *Arabidopsis*, which is involved in abiotic stress and plant development (Park et al. 2014). Furthermore, AS can generate potential isoforms of miRNAs originated from intronic regions (like miR788, miR862, miR853, miR848, miR844, miR834, and miR166a), indicating that these intronic-miRNAs may have significant involvement in plant thermotolerance mechanism (Mishra and Bohra 2018).

Epigenetic mechanisms, including DNA methylation, are vital for plant growth, development, and survival under adverse environmental conditions. This phenomenon has primary importance in gene expression regulation to shape a plant's fate (Rakei et al. 2015). Methylated forms of *Nt-EXPA5* and *CycD3-1* exhibit altered expression in tobacco plants facing heat stress (Centomani et al. 2015). In the same way, the expression of miRNA and respective target genes might be affected by DNA methylation under heat stress. In *Populus simonii*, some miRNAs (miR156i, miR167h, miR396a, miR396h, miR396g) are methylated under cold and heat stress at CG and CNG sites, respectively. In another study, it was observed that heat stress increased the expression of miR390c and decreased the expression of target *IDH* (isocitrate-dehydrogenase) gene, which is involved in peroxisomes biosynthesis. Likewise, in response to heat stress, the ^{m}CNG-modified Ptc-miR156i and Ptc-miR156j expression were induced compared to a down-regulation of target genes (*LPCAT1* and 2) that play a role in metabolism of lipids and glycerol-phospholipids. Moreover, heat stress increased expression of miR396e and miR396g in contrast to target genes (*ACOX1* and *ACOX2*), which are involved in fatty acid degradation and alpha linolenic acid metabolism (Ci et al. 2015; Biswas and Mano 2015).These reports indicate that in response to high temperature stress methylation of DNA may impact the expression of miRNAs and target genes, potentially using miRNA-mediated gene-silencing mechanisms (Song et al. 2015; Ci et al. 2015).

8.3 Small Interfering RNA (siRNA)

siRNAs are endogenous RNAs, approximately 21–24 nt in length, derived from DCL family (DCL1), that catalyze the processing of double-stranded RNA precursors (Axtell 2013). siRNAs can be further classified into two groups on the basis of their biogenesis and function; heterochromatic siRNAs or natural antisense transcript siRNAs (nat-siRNAs) and trans acting siRNAs (ta-siRNAs) (Axtell 2013; Sunkar and Zhu 2007). Sunkar and Zhu (2007) reveal that siRNAs help the plants to combat abiotic stress conditions. Afterwards, it was observed that nat-siRNAs (products of natural *cis*-antisense pairing of *P5CDH* and *SRO5* transcripts) can modulate salt stress tolerance in *Arabidopsis*. Unfortunately, it is not known how siRNAs can help plants to survive high temperature stress. Yao et al. (2010) report that in response to cold or heat stress, the expression of three siRNAs is either up-regulated or down-regulated, respectively. In heat-treated *Arabidopsis* seedlings, a copia-type transposon (*ONSEN*) is activated. Meanwhile, this transposon is even more actively stimulated in siRNA biogenesis-impaired mutant plants under heat stress. It suggests that siRNAs are involved in the suppression of *ONSEN* transcripts (Ito et al. 2011). It has been observed that the progeny of heat-stressed siRNA-deficient plants carry a higher frequency of new *ONSEN* insertions. *trans*-Acting siRNAs or ta-siRNAs originate from miRNA-mediated transcript processing of *TSA* gene. These siRNAs function in a way similar to miRNAs (Axtell 2013). In *Arabidopsis*, miR173 is cleaved to produce ta-siRNAs, which target *HTT1* and *HTT2* (heat-induced TAS 1 target 1 and 2) genes. Both *HTT1* and *HTT2* are significantly induced under heat stress conditions and play a role in thermos tolerance (Khraiwesh et al. 2012; M. Y. Li et al. 2014; S. X. Li et al. 2014). It was observed that plants overexpressing *HTT1* and *HTT2* also accumulate higher levels of various HSF genes to enhance heat-stress tolerance. Interestingly, the HTT genes were also activated in mutants overexpressing HsfA1a in elevated temperatures. In contrast, the over-expression of *TAS1a* enhanced the accumulation of TAS1-siRNAs, reduced the expression of

HTT genes, and plants became more sensitive to heat stress. It suggested that *TAS1a* is involved in negative regulation HTT genes and thermos tolerance. nAT-siRNA represents another group of siRNAs, which originate through hybridization of two individually transcribed RNAs (Axtell 2013). It was noticed in *B. rapa* that nat-siRNAs (cleavage products of 12 *cis*-NATs) positively respond to elevated temperatures and exhibit strand biasness. From the same loci, only those transcripts were up-regulated that generated heat-responsive nat-siRNAs, and transcripts from complementary strands were down-regulated (Yu et al. 2013).

8.4 Long Noncoding RNAs (lncRNAs)

The remarkable growth of high-throughput sequencing technologies and next-generation sequencing (NGS) methodologies has revolutionized the identification and characterization of novel ncRNAs. In plants, lncRNAs represent another class of functional RNAs, which are generally more than 200 nt in length (Rai et al. 2018). There are two systems for the classification of lncRNAs. According to the position in genomic DNA, there are two subgroups of lncRNAs as intronic lncRNAs and antisense lncRNA (Wierzbicki 2012). According to function, lncRNAs act as (i) *cis*-acting lncRNAs, (ii) target mimics, (iii) scaffold and guides and (iv) mediators of RNA-dependent DNA methylation (Rai et al. 2018). The lncRNAs have been recognized among key transcriptional regulators, which are involved in a range of crucial biological processes, from epigenetic to post-transcriptional levels of gene expression regulation (Heo and Sung 2011). Various studies summarize the importance of lncRNAs in plant responses to biotic and abiotic stresses (Nejat and Mantri 2018; Deng et al. 2018; Meng et al. 2019). However, our understanding is still limited on mechanisms that lncRNAs use to perform their functions. Various lncRNAs have been functionally characterized in plant stress-responsive pathways. It was reported that ln-173 (lncRNA) was not activated by heat stress as compared to its target gene *SUS4* (sucrose synthase 4) (Di et al. 2014). Xin et al. (2011) use microarray and a sequencing-by-synthesis (SBS) approach to identify 125 putative lncRNAs in wheat in response to fungal infection and heat stress. In *Arabidopsis*, Di et al. (2014) identified that 58 poly(A)- and 245 poly(A)+ lncRNAs were differentially regulated under stress conditions. However, application of heat stress neutralized most of the differential expressions. Similarly, in *P. simonii*, a lncRNA (PsiLncRNA00268512) showed differential expression under heat stress (Song et al. 2015). Likewise, 34 lncRNAs were found to be particularly expressed under heat stress conditions in *B. rapa*. It was also observed that the majority of respective target genes belonged to heat-stress-responsive genes (Song et al. 2016). The use of bioinformatics and ever-developing high-throughput sequencing technologies has significantly assisted the identification of novel lncRNAs.

8.5 Circular RNAs (circRNAs)

Circular RNAs (circRNAs) are covalently closed and single-stranded transcripts (for details see Chu et al. 2018). These RNAs have been identified both in prokaryotes and eukaryotes. Initially, circRNAs were considered artifacts of splicing events. However, now circRNAs have been identified to play regulatory roles in biological function transcription, AS, binding of miRNAs or proteins, and stress response (Lai et al. 2018; Litholdo Jr. and da Fonseca 2018). The first circRNA was discovered in *Arabidopsis* in 2014, but still a knowledge gap is present regarding circRNAs in plants. Ye et al. (2015) performed a genome-wide analysis and demonstrated that circRNAs were substantially present in *Arabidopsis* and *Oryza sativa*. Recently, the role of circRNAs has been investigated in plant stress responses (Zhang et al. 2019). In response to heat stress, 1583 circRNAs were differentially expressed in *Arabidopsis* (Pan et al. 2018). Under heat stress conditions, the length of circRNAs is increased probably because of an increased number of exons involved in circularization, suggesting that heat stress affects not only the quantity but also the quality (alternate back splicing and circularization) of circRNAs. However, a great deal of research is still required to reveal the role of circRNAs in plants. It is anticipated that NGS technologies coupled with computational strategies could play crucial roles in deciphering the biological role of circRNAs in plant response to heat stress.

8.6 Conclusions and Future Perspectives

Heat stress has become a major limiting factor for crop production worldwide, because it adversely affects plant growth, development, and yield. Thus, plants have evolved various adaptive mechanism to cope with heat stress. Plant ncRNAs play an important role in heat response by ncRNA-target pathway in heat response networks of plants. However, ncRNA research is attaining a central position in gene regulation studies to improve crop performance. There is a lot to improve in ncRNA nomenclature, functional analysis of miRNAs, ncRNA spatiotemporal expression, integrated regulation of ncRNA-target gene network, and newly identified classes of circRNAs. Moreover, most plant ncRNAs have been reported in a limited range of species under heat stress conditions, and their functional relevance requires comprehensive research in plants adapted to extreme temperatures.

References

Ali, M.A., Azeem, F., Li, H., and Bohlmann, H. (2017). Smart parasitic nematodes use multifaceted strategies to parasitize plants. *Frontiers in Plant Science* 8: 1699–1710.

Ali, M.A., Anjam, M., Nawaz, M.A. et al. (2018). Signal transduction in plant–nematode interactions. *International Journal of Molecular Sciences* 19: 1648.

Axtell, M.J. (2013). Classification and comparison of small RNAs from plants. *Annual Review of Plant Biology* 64: 137–159.

Banerjee, S., Sirohi, A., Ansari, A.A., and Gill, S.S. (2017). Role of small RNAs in abiotic stress responses in plants plant gene role of small RNAs in abiotic stress responses in plants. *Plant Gene* 180: 189.

Barku, M.M., Bashasab, F., Sudipta, G., and Krishnaraj, P.U. (2013). RNA mediated *in situ* hybridization of miR171 and miR397a in leaf and ambient root tissues revealed expressional homogeneity in response to shoot heat shock in *Arabidopsis thaliana*. *Journal of Plant Biochemistry and Biotechnology* 23: 93–103.

Biswas, M.S. and Mano, J.I. (2015). Lipid peroxide-derived short-chain carbonyls mediate H2O2-induced and

NaCl-induced programmed cell death in plants. *Plant Physiology* 168: 885–898.

Borges, F. and Martienssen, R.A. (2015). *The Expanding World of Small RNAs in Plants*. Nature Publishing Group.

Bulgakov, V.P. and Avramenko, T.V. (2015). New opportunities for the regulation of secondary metabolism in plants: focus on microRNAs. *Biotechnology Letters* 37 (9): 1719–1727.

Centomani, I., Sgobba, A., D'Addabbo, P. et al. (2015). Involvement of DNA methylation in the control of cell growth during heat stress in tobacco BY-2 cells. *Protoplasma* 252: 1–9.

Chen, L., Ren, Y.Y., Zhang, V.L. et al. (2012). Genome-wide identification and expression analysis of heat responsive and novel microRNAs in *Populus tomentosa*. *Gene* 504: 160–165.

Chu, Q., Bai, P., Zhu, X. et al. (2018). https://doi.org/10.1093/bib/bby111). Characteristics of plant circular RNAs. *Briefings in Bioinformatics*.

Cui, L.G., Shan, J.X., Shi, M. et al. (2014). The *miR156-SPL9- DFR* pathway coordinates the relationship between development and abiotic stress tolerance in plants. *The Plant Journal* 80: 1108–1117.

Deng, P., Liu, S., Nie, X. et al. (2018). Conservation analysis of long non-coding RNAs in plants. *Science China Life Sciences* 61 (2): 190–198.

Di, C., Yuan, J.P., Wu, Y. et al. (2014). Characterization of stress-responsive lncRNAs in *Arabidopsis thaliana* by integrating expression, epigenetic and structural features. *The Plant Journal* 80: 848–861.

Gallavotti, A. (2013). The role of auxin in shaping shoot architecture. *Journal of Experimental Botany* 64 (9): 2593–2608.

Giacomelli, J.I., Weigel, D., Chan, R.L., and Manavella, P.A. (2012). Role of recently evolved miRNA regulation of sunflower HaWRKY6 in response to temperature damage. *The New Phytologist* 195: 766–773.

Guan, Q.M., Lu, X.Y., Zeng, H.T. et al. (2013). Heat stress induction of miR398 triggers a regulatory loop that is critical for thermotolerance in *Arabidopsis*. *The Plant Journal* 74: 840–851.

Guerra, D., Crosatti, C., Khoshro, H.H. et al. (2015). Post-transcriptional and post-translational regulations microRNAs involved in temperature stress response. *BMC Genomics* 15: 242.

Guo, H.S., Xie, Q., Fei, J.F., and Chua, N.H. (2005). MicroRNA directs mRNA cleavage of the transcription factor NAC1 to downregulate auxin signals for *Arabidopsis* lateral root development. *Plant Cell* 17: 1376–1386.

Heo, J.B. and Sung, S. (2011). Vernalization-mediated epigenetic silencing by a long intronic noncoding RNA. *Science* 331: 76–79.

Hivrale, V., Yun, Z., Puli, C.O.R. et al. (2016). Characterization of drought- and heat-responsive microRNAs in switchgrass. *International Journal of Plant Sciences* 242: 214–223.

Huang, B. and Xu, C.P. (2008). Identification and characterization of proteins associated with plant tolerance to heat stress. *Journal of Integrative Plant Biology* 50: 1230–1237.

Ito, H., Gaubert, H., Bucher, E. et al. (2011). An siRNA pathway prevents transgenerational retrotransposition in plants subjected to stress. *Nature* 472: 115–119.

Kamiya, N., Itoh, J., Morikami, A. et al. (2003). The SCARECROW gene's role in asymmetric cell divisions in rice plants. *The Plant Journal* 36: 45–54.

Khaksefidi, R.E., Mirlohi, S., Khalaji, F. et al. (2015). Differential expression of seven conserved microRNAs in response to abiotic stress and their regulatory network in *Helianthus annuus*. *Frontiers in Plant Science* 6: 741.

Khraiwesh, B., Zhu, J.K., and Zhu, J. (2012). Role of miRNAs and siRNAs in biotic and abiotic stress responses of plants. *Biochimica et Biophysica Acta* 1819: 137–148.

Kim, J.J., Lee, J.H., Kim, W. et al. (2012). The *miR156-SPL3* module regulates ambient temperature-responsive flowering via *FT* in *Arabidopsis thaliana*. *Plant Physiology* 159: 461–478.

Kotak, S., Larkindale, J., Lee, U. et al. (2007). Complexity of the heat stress response in plants. *Current Opinion in Plant Biology* 10: 310–316.

Kruszka, K., Pacak, A., Swidabarteczka, A. et al. (2014). Transcriptionally and post-transcriptionally regulated microRNAs in heat stress response in barley. *Journal of Experimental Botany* 65: 6123–6135.

Kumar, S. (2018). Epigenomics of plant responses to environmental stress. *Epigenomes* 2 (1): 6.

Kumar, R.R., Pathak, H., Sharma, S.K. et al. (2015). Novel and conserved heat-responsive microRNAs in wheat (*Triticum aestivum*). *Functional and Integrative Genomics* 15: 1–26.

Lai, X., Bazin, J., Webb, S. et al. (2018). CircRNAs in plants. In: *Circular RNAs*, 329–343. Singapore: Springer.

Li, M.Y., Wang, F., Xu, Z.S. et al. (2014). High throughput sequencing of two celery varieties small RNAs identifies microRNAs involved in temperature stress response. *BMC Genomics* 15: 242.

Li, S.X., Liu, J.X., Liu, Z.Y. et al. (2014). Heat-induced tas1 target1 mediates thermotolerance via heat stress transcription factor A1a-directed pathways in *Arabidopsis*. *Plant Cell* 26: 1764–1780.

de Lima, J.C., Loss-Morais, G., and Margis, R. (2012). microRNAs play critical roles during plant development and in response to abiotic stresses. *Genetics and Molecular Biology* 35: 1069–1077.

Lin, J.S., Kuo, C.C., Yang, I.C. et al. (2018). MicroRNA160 modulates plant development and heat shock protein gene expression to mediate heat tolerance in *Arabidopsis*. *Frontiers in Plant Science* 9: 68.

Litholdo, C.G. and da Fonseca, G.C. (eds.) (2018). Circular RNAs and plant stress responses. In: *Circular RNAs*, 345–353. Singapore: Springer.

Liu, F.L., Wang, W.J., Sun, X.T. et al. (2014). RNA-Seq revealed complex response to heat stress on transcriptomic level in *Saccharina japonica* (Laminariales, Phaeophyta). *Journal of Applied Phycology* 26: 1585–1596.

Liu, Q., Yang, T., Yu, T. et al. (2017). Integrating small RNA sequencing with QTL mapping for identification of miRNAs and their target genes associated with heat tolerance at the flowering stage in rice. *Frontiers in Plant Science* 8: 43.

Liu, H., Yu, H., Tang, G., and Huang, T. (2018). Small but powerful: function of microRNAs in plant development. *Plant Cell Reports* 37 (3): 515.

Lu, S.F., Sun, Y.H., and Chiang, V.L. (2008). Stress-responsive microRNAs in *Populus*. *The Plant Journal* 55: 131–151.

Lu, X., Guan, Q., and Zhu, J. (2013). Downregulation of CSD2 by a heat-inducible miR398 is required for thermotolerance in *Arabidopsis*. *Plant Signaling & Behavior* 8: e24952.

Mahale, B.M., Fakrudin, B., Ghosh, S., and Krishnaraj, P.U. (2013). LNA mediated in situ hybridization of miR171 and miR397a in leaf and ambient root tissues revealed expressional homogeneity in response to shoot heat shock in *Arabidopsis thaliana*. *Journal of Plant Biochemistry and Biotechnology* 23 (1): 93–103.

Matsukura, S., Mizoi, J., Yoshida, T. et al. (2010). Comprehensive analysis of rice *DREB2*-type genes that encode transcription factors involved in the expression of abiotic stress-responsive genes. *Molecular Genetics and Genomics* 283: 185–196.

May, P., Liao, W., Wu, Y. et al. (2013). The effects of carbon dioxide and temperature on microRNA expression in *Arabidopsis* development. *Nature Communications* 4: 405–415.

Megha, S., Basu, U., and Kav, N.N.V. (2018). Regulation of low temperature stress in plants by microRNAs. *Plant, Cell & Environment* 41: 1–15.

Meng, Y., Chen, D., Ma, X. et al. (2010). Mechanisms of microRNA-mediated auxin signaling inferred from the rice mutant *osaxr*. *Plant Signaling & Behavior* 5: 252–254.

Meng, Y., Xing, L., Li, K. et al. (2019). Genome-wide identification, characterization and expression analysis of novel long non-coding RNAs that mediate IBA-induced adventitious root formation in apple rootstocks. *Plant Growth Regulation* 87 (2): 287–302.

Mishra, A. and Bohra, A. (2018). Non-coding RNAs and plant male sterility: current knowledge and future prospects. *Plant Cell Reports* 37 (2): 177–191.

Nejat, N. and Mantri, N. (2018). Emerging roles of long non-coding RNAs in plant response to biotic and abiotic stresses. *Critical Reviews in Biotechnology* 38 (1): 93–105.

Noman, A., Fahad, S., Aqeel, M. et al. (2017). miRNAs: major modulators for crop growth and development under abiotic stresses. *Biotechnology Letters* 39 (5): 685–700.

Ohama, N., Sato, H., Shinozaki, K., and Yamaguchi-shinozaki, K. (2017). Transcriptional regulatory network of plant heat stress response. *Trends in Plant Science* 22 (1): 53–65.

Pan, T., Sun, X., Liu, Y. et al. (2018). Heat stress alters genome-wide profiles of circular RNAs in *Arabidopsis*. *Plant Molecular Biology* 96 (3): 217–229.

Park, Y.J., Lee, H.J., Kwak, K.J. et al. (2014). MicroRNA400-guided cleavage of pentatricopeptide repeat protein mRNAs renders *Arabidopsis thaliana* more susceptible to pathogenic bacteria and fungi. *Plant & Cell Physiology* 55: 1660–1668.

Prasad, P.V.V. and Jagadish, S.V.K. (2015). Field crops and the fear of heat stress: opportunities, challenges and future directions. *Procedia Environmental Sciences* 29: 36–37.

Qin, F., Kakimoto, M., Sakuma, Y. et al. (2007). Regulation and functional analysis of ZmDREB2A in response to drought and heat stresses in *Zea mays* L. *The Plant Journal* 50: 54–69.

Rai, M.I., Alam, M., Lightfoot, D.A. et al. (2018). Classification and experimental identification of plant long non-coding RNAs. *Genomics*.

Rakei, A., Maali-Amiri, R., Zeinali, H., and Ranjbar, M. (2015). DNA methylation and physio-biochemical analysis of chickpea in response to cold stress. *Protoplasma* 253: 1–16.

Reyes, J.L. and Chua, N.H. (2007). ABA induction of miR159 controls transcript levels of two MYB factors during *Arabidopsis* seed germination. *The Plant Journal* 49: 592–606.

Rhoades, M.W., Reinhart, B.J., Lim, L.P. et al. (2002). Prediction of plant microRNA targets. *Cell* 110: 513–520.

Sailaja, B., Voleti, S.R., Subrahmanyam, D., and Sarla, N. (2013). Prediction and expression analysis of miRNAs associated with heat stress in rice (*Oryza sativa*). *Journal of Experimental Botany* 23: 13–25.

Schommer, C., Palatnik, J.F., Aggarwal, P. et al. (2008). Control of jasmonate biosynthesis and senescence by miR319 targets. *PLoS Biology* 6: e230.

Shang, X., Cao, Y., and Ma, L. (2017). Alternative splicing in plant genes: a means of regulating the environmental fitness of plants. *International Journal of Molecular Sciences* 18 (2): 432–441.

Shriram, V., Kumar, V., Devarumath, R.M. et al. (2016). MicroRNAs as potential targets for abiotic stress tolerance in plants. *Frontiers in Plant Science* 7: 817–826.

Smoczynska, A., Sega, P., Stepien, A. et al. (2019). miRNA detection by stem-loop RT-qPCR in studying microRNA biogenesis and microRNA responsiveness to heat stress. *Plant MicroRNAs* 131: 150.

Song, Y., Ci, D., Tian, M., and Zhang, D.Q. (2015). Stable methylation of a noncoding RNA gene regulates gene expression in response to abiotic stress in *Populus simonii*. *Journal of Experimental Botany* 67: 1477–1492.

Song, X., Liu, G., Huang, Z. et al. (2016). Temperature expression patterns of genes and their coexpression with lncRNAs revealed by RNA-Seq in non-heading Chinese cabbage. *BMC Genomics* 17: 297.

Stief, A., Altmann, S., Hoffmann, K. et al. (2014). *Arabidopsis* miR156 regulates tolerance to recurring environmental stress through *SPL* transcription factors. *Plant Cell* 26: 1792–1807.

Sunkar, R. and Zhu, J.K. (2007). Micro RNAs and short-interfering RNAs in plants. *Journal of Integrative Plant Biology* 49: 817–826.

Vaucheret, H. (2009). *AGO1* homeostasis involves differential production of 21-nt and 22-nt miR168 species by MIR168a and MIR168b. *PLoS One* 4: e6442.

Wang, Y., Sun, F., Cao, H. et al. (2012). TamiR159 directed wheat TaGAMYB cleavage and its involvement in another development and heat response. *PLoS One* 7: e48445.

Wierzbicki, A.T. (2012). The role of long non-coding RNA in transcriptional gene silencing. *Current Opinion in Plant Biology* 15: 517–522.

de Wit, M., Lorrain, S., and Fankhauser, C. (2014). Auxin-mediated plant architectural changes in response to shade and high temperature. *Physiologia Plantarum* 151 (1): 13–24.

Wu, G., Park, M.Y., Conway, S.R. et al. (2009). The sequential action of miR156 and miR172 regulates developmental timing in *Arabidopsis*. *Cell* 138: 750–759.

Xin, M., Yu, W., Yao, Y. et al. (2010). Diverse set of microRNAs are responsive to powdery mildew infection and heat stress in wheat (*Triticum aestivum* L.). *BMC Plant Biology* 10: 123–110.

Xin, M., Wang, Y., Yao, Y. et al. (2011). Identification and characterization of wheat long non-protein coding RNAs responsive to powdery mildew infection and heat stress by using microarray analysis and SBS sequencing. *BMC Plant Biology* 11: 61–10.

Yan, K., Liu, P., Wu, C.A. et al. (2012). Stress-induced alternative splicing provides a mechanism for the regulation of microRNA processing in *Arabidopsis thaliana*. *Molecular Cell* 48: 521–531.

Yao, Y., Ni, Z., Peng, H. et al. (2010). Non-coding small RNAs responsive to abiotic stress in wheat (*Triticum aestivum* L.). *Functional & Integrative Genomics* 10: 187–190.

Ye, C.Y., Chen, L., and Liu, C. (2015). Widespread noncoding circular RNAs in plants. *The New Phytologist* 208: 88–95.

Yu, X., Wang, H., Lu, Y.Z. et al. (2012). Identification of conserved and novel microRNAs that are responsive to heat stress in *Brassica rapa*. *Journal of Experimental Botany* 63: 1025–1038.

Yu, X., Yang, J., Li, X. et al. (2013). Global analysis of cisnatural antisense transcripts and their heat-responsive nat-siRNAs in *Brassica rapa*. *BMC Plant Biology* 13: 208.

Zhang, B. (2015). MicroRNA: a new target for improving plant tolerance to abiotic stress. *Journal of Experimental Botany* 66 (7): 1749–1761.

Zhang, P., Fan, Y., Sun, X. et al. (2019). A large-scale circular RNA profiling reveals universal molecular mechanisms responsive to drought stress in maize and *Arabidopsis*. *The Plant Journal*.

Zhao, J., He, Q., Chen, G. et al. (2016). Regulation of non-coding RNAs in heat stress responses of plants. *Journal of Experimental Botany* 7: 1–9.

Zhao, J., Yuan, S., Zhou, M. et al. (2019). Transgenic creeping bentgrass overexpressing Osa-miR393a exhibits altered plant development and improved heat stress tolerance. *Plant Biotechnology Journal* 17 (1): 233–251.

Zhu, C., Ding, Y.F., and Liu, H.L. (2011). MiR398 and plant stress responses. *Physiologia Plantarum* 143: 1–9.

Zhu, H., Zhang, Y., Tang, R. et al. (2019). Banana sRNAome and degradome identify microRNAs functioning in differential responses to temperature stress. *BMC Genomics* 20 (1): 33.

9

Heat Shock Proteins: Master Players for Heat-stress Tolerance in Plants during Climate Change

Annu Yadav[1], *Jitender Singh*[1], *Koushlesh Ranjan*[2], *Pankaj Kumar*[1], *Shivani Khanna*[1], *Madhuri Gupta*[1], *Vinay Kumar*[3], *Shabir Hussain Wani*[4], *and Anil Sirohi*[1]

[1] *College of Biotechnology, Sardar Vallabhbhai Patel University of Agriculture & Technology, Meerut, Uttar Pradesh, India*
[2] *Department of Veterinary Physiology and Biochemistry, College of Veterinary and Animal Sciences, Sardar Vallabhbhai Patel University of Agriculture and Technology, Meerut, Uttar Pradesh, India*
[3] *Department of Biotechnology, Modern College of Arts, Science and Commerce, Savitribai Phule Pune University, Pune, India*
[4] *Mountain Research Centre for Field Crops, Khudwani, Sher-e-Kashmir University of Agricultural Sciences and Technology of Kashmir, Jammu and Kashmir, India*

9.1 Introduction

Climate change and constantly raising ambient temperature are considered two of the most detrimental stress factors for plant species. The global temperature is predicted to rise between 0.28 and 0.77 °C during 2016–2020 as per report of Met office UK (www.metoffice.gov.uk/research/climate/seasonal-to-decadal/long-range/decadal-fc). High-temperature is one of the major abiotic stresses limiting plant distribution, growth, and productivity (Figure 9.1). At the molecular level, one of the best-characterized environmental responses is the response to high temperature (HT) or heat shock. Plant species including crop plants such as soybean, pea, maize, and wheat, etc., adapt to high temperate environments and begin to synthesize heat shock proteins (HSPs) when tissue temperatures exceed 32–40 °C.

Heat Stress Tolerance in Plants: Physiological, Molecular and Genetic Perspectives, First Edition.
Edited by Shabir Hussain Wani and Vinay Kumar.

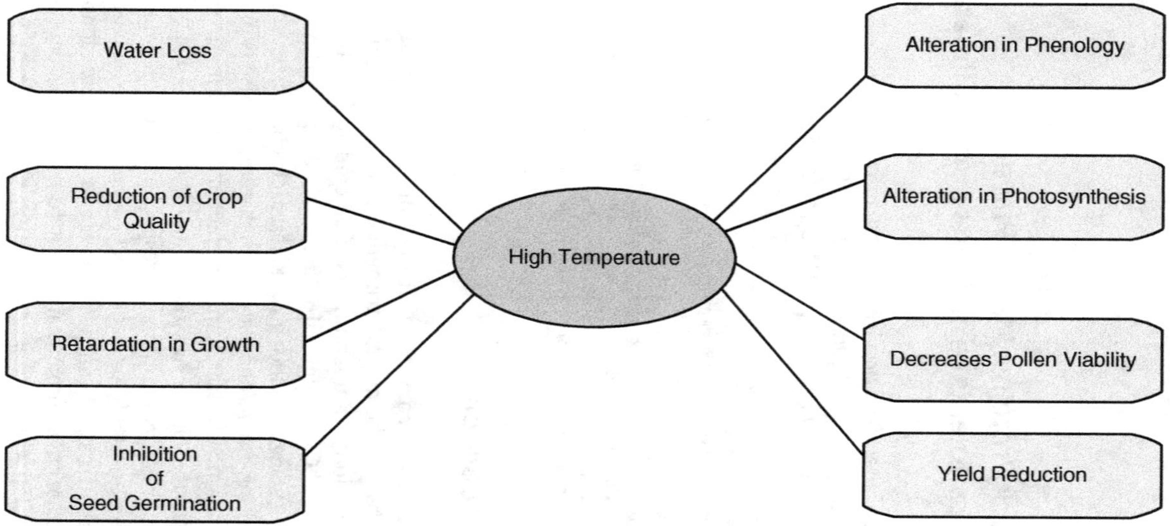

Figure 9.1 Major effects of high temperature on plants.

HSP genes are present under normal conditions, but are expressed only when plants (as well as all other organisms) are exposed to HT stress. HSPs originally were discovered when it was observed that heat shock produced chromosomal puffs in the salivary glands of fruit flies (*Drosophila*). When plants are exposed to environmental stress like heat, they respond with changes in physiological and gene expression patterns (Molinier et al. 2006). Expression of HSPs is an important adaptive strategy for stress tolerance (Feder and Hofmann 1999). However, most of the stress-induced signals and reactions are temporary and reset to the basal level once the stress is relieved. However, it may be stable and as a form of "stress memory" (Chinnusamy and Zhu 2009). HSPs can play a crucial role in protecting plants against stress by reestablishing normal protein conformation and thus cellular homeostasis (Wang et al. 2004). Heat shock transcription factors (HSFs) are the transcriptional activators of HSPs (Banti et al. 2010) and play an important role in the synthesis of HSPs.

Massive production of HSPs in plants is an important characteristic responsible for the acquisition of thermotolerance. Thermotolerance in higher plants is correlated with LMW-HSP (HSP18.1, HSP17.9) expression in wheat (Basha et al. 2004) and HMW–HSP (HSP101) expression in *Arabidopsis* (Queitsch et al. 2000). However, study of *Arabidopsis* mutant phenotypes suggests that high accumulation of HSPs is only one critical aspect of acquired thermotolerance (Larkindale et al. 2005). Apart from HSPs, several cellular signaling factors including abscisic acid (ABA), salicylic acid, ethylene, AOS, antioxidants, the ubiquitin pathway, and membrane fatty acids also play crucial roles in thermotolerance (Larkindale et al. 2005). Wang and Luthe (2003) report that *LMW-HSP* genes such as *ApHsp26.2* and *ApHsp26.7a* were found highly accumulated in heat-tolerant creeping bentgrass variant than heat-sensitive ones. Kumar et al. (2015) identified HSP gene (VfHsp17.9-CII) expression in faba bean at HT 30–40 °C, which protects physiology, photosynthetic assimilation rate, and changes in membrane characteristics. In this chapter we summarize comprehensive information and recent findings of HSPs in plants and their role under stress conditions.

9.2 Classification of HSPs

HSPs are generally categorized into five evolutionarily conserved families such as HSP100 (OR ClpB), HSP90, HSP70, (OR DnaK), HSP60 (Gro E) and sHSPs, (small heat stress proteins, < 40 kDa) (Table 9.1).

9.2.1 HSP100

The family of Hsp100 belongs to super family AAA ATPase and act as molecular chaperones (Bukau et al. 2006). Hsp100 family is expressed under conditions of not only heat stress but also high salt, desiccation, ABA, and cold stress. ClpA (caseinolytic protease or Clp) was the first protein of Hsp100 class that was discovered as a component of adenosine diphosphate (ATP) dependent protease, which can solubilize and reactivate almost any protein that becomes aggregated after severe stress (Gottesman et al. 1990). The role of HSP100 as a stress tolerant has been studied in several plant species such as *Arabidopsis* (*Arabidopsis thaliana*), tobacco (*Nicotiana tabacum*), soybean (*Glycine max*), lima bean (*Phaseolus lunatus*), maize (*Zea mays*), wheat (*Triticum sativum*), and rice (*Oryza sativa*) (Wang et al. 2004).

9.2.1.1 Structure of HSP100

Hsp100 class of chaperones belongs to the AAA+ superfamily of ATPases having ~200–250 amino acids which comprise α-helical domain and a walker-type nucleotide-binding domain with glycine, lysine, threonine, and serine residues for ATP binding (Bukau et al. 2006). Hsp100 is divided into t: class I and class II chaperones. Class I includes Hsp104, bacterial ClpB, ClpA, and ClpC, whereas class II includes ClpX and HslU. Clp/Hsp100 proteins assemble into ring structures and have six protomers (Doyle and Wickner 2009) (Figure 9.2a).

9.2.1.2 Mode of Action: The HSP100 Chaperone Cycle

Hsp100 class of chaperones such as Hsp104 and ClpB (class I family) contains binding domains which have walker-like domain A and B along with sensor 1 and 2 motifs. Hsp100 works in oligomeric form along with the ATP and HSP70/Dnak (Glover and Lindquist 1998). ATP hydrolyzes into adenosine

Table 9.1 List of heat shock proteins including their localization and functions.

Classes	Members	Localization	Functions	References
Hsp100/Clp	Hsp100, Subfamily: Class I (ClpB, ClpA/C, ClpD) Class II (ClpM, ClpN, ClpX, ClpY)	Cytosol, mitochondria, chloroplast	Disaggregation, unfolding/facilitate reactivation of proteins denatured by heat	Queitsch et al. 2000; Fender and Connell 1990; Campbell et al. 2001
Hsp90	Hsp90	Cytosol, chloroplast, mitochondria, endoplasmic reticulum	Facilitating maturation of signaling molecules, genetic buffering Act as part of a multi-chaperone machine together with HSP70 and cooperates with a cohort of co-chaperone	Fender and Connell 1990; Usman et al. 2014; Wang et al. 2004
Hsp70		Cytosol, chloroplast, mitochondria, endoplasmic reticulum	Preventing aggregation, assisting refolding, protein import and translocation, signal transduction, and transcriptional activation	Wang et al. 2004; Usman et al. 2014
Hsp60 or Chaperonin	Subfamily: Group I and Group II	Chloroplast, mitochondria, cytosol	Folding and assisting refolding	Boston et al. 1996; Usman et al. 2014; Wang et al. 2004
Small Hsp	<40 kDa like Hsp17.6, Hsp17.9, Hsp21, Hsp26.2, Hsp23, Hsp23.3, Hsp40	Cytosol, chloroplast, mitochondria, endoplasmic reticulum	Aggregation preventing, stabilizing non-native proteins, work in association with ATP-depended HSP like HSP70	Usman et al. 2014; Zhang and Denlinger 2010; Wang et al. 2004

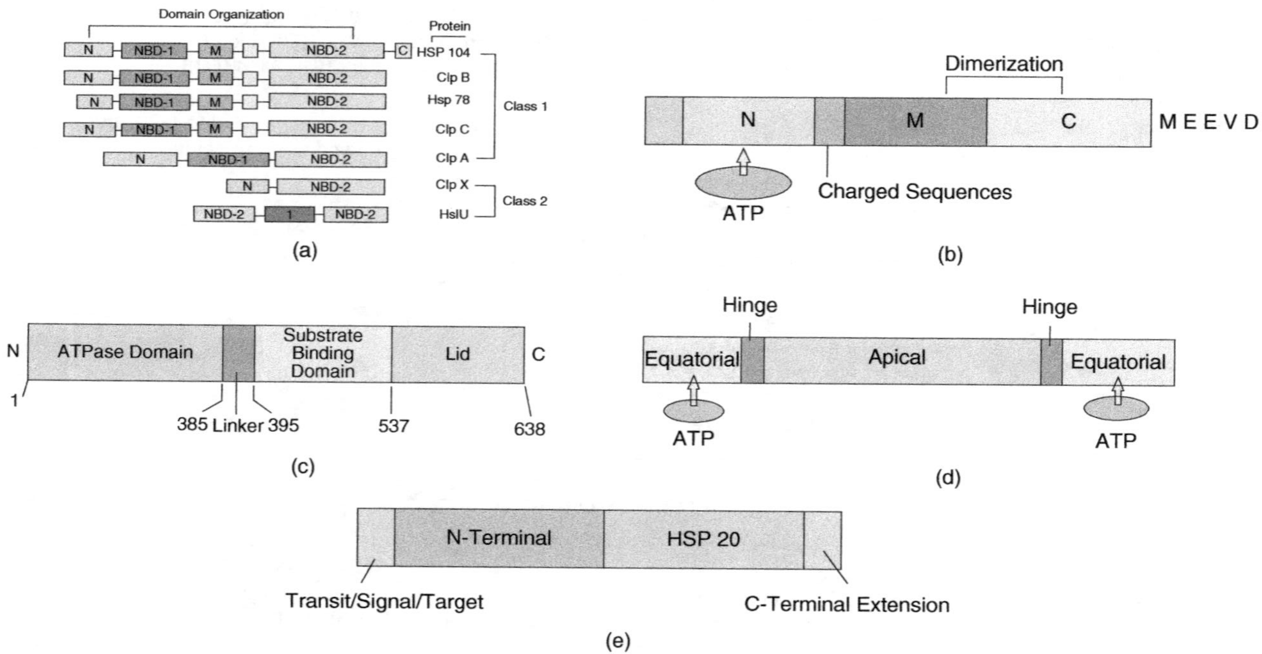

Figure 9.2 Domain organization of (a) HSP100 (class I and class II); (b) HSP90; (c) HSP70; (d) HSP60; and (e) small heat shock proteins.

diphosphate (ADP) and releases energy which helps in protein disaggregation (Glover and Lindquist 1998). HSP70/Dnak involves in the protein disaggregation by binding with hsp104/clpB. HSP70 may work in two ways. First, it extracts the protein from disaggregate to aggregate and also is involved in ATP hydrolysis. Second, it works as a reactivator of protein if protein fails to remodel through HSP104/ClpB. After the activity of ATP hydrolysis, HSP70 releases native form of protein (Figure 9.3a).

Class II proteins such as clpX and HslU are a ATP dependent and with protein unfolding and proteolytic activity. ClpX is involved in (i) binding of ATP to clpX which causes the conformational changes due to the hydrolysis of ATP and binds with the misfolded form of protein or substrate, (ii) conversion of misfolded protein into stable three-dimensional protein, and (iii) translocation and degradation of stable three-dimensional protein structures (Lund 2001; Parsell and Lindquist 1993; Bukau et al. 2006) (Figure 9.3b).

9.2.2 HSP90

HSP90 is a molecular chaperone found in bacteria and eukaryotes but absent in archaebacteria. An abundant cytosolic protein in eukaryotes acts as part of a multi-chaperone machine together with HSP70 and cooperates with co-chaperons, including Hip (HSP70 interacting protein), Hop (HSP70/HSP90 organizing protein), p23 and HSP40 (Ludwig-Muller et al. 2000). HSP90 genes have been identified from many plants under various stress conditions such as changes in temperature, salinity, and heavy metals.

Recent research shows that Hsp90 together with Hsp70 and co-chaperones plays an important role in stabilizing and activating homeostasis of protein (i.e. proteostasis), protein trafficking, transcriptional regulation, chromatin remodeling, disease resistance, and DNA repair (Wang et al. 2004; Xu et al. 2012). It also regulates diverse cellular functions and exerts marked effects on normal biology, disease, and evolutionary processes (Taipale et al. 2010). Under normal conditions, HSP90.2 in the cytoplasm suppresses the HSF and negatively regulates the transcription

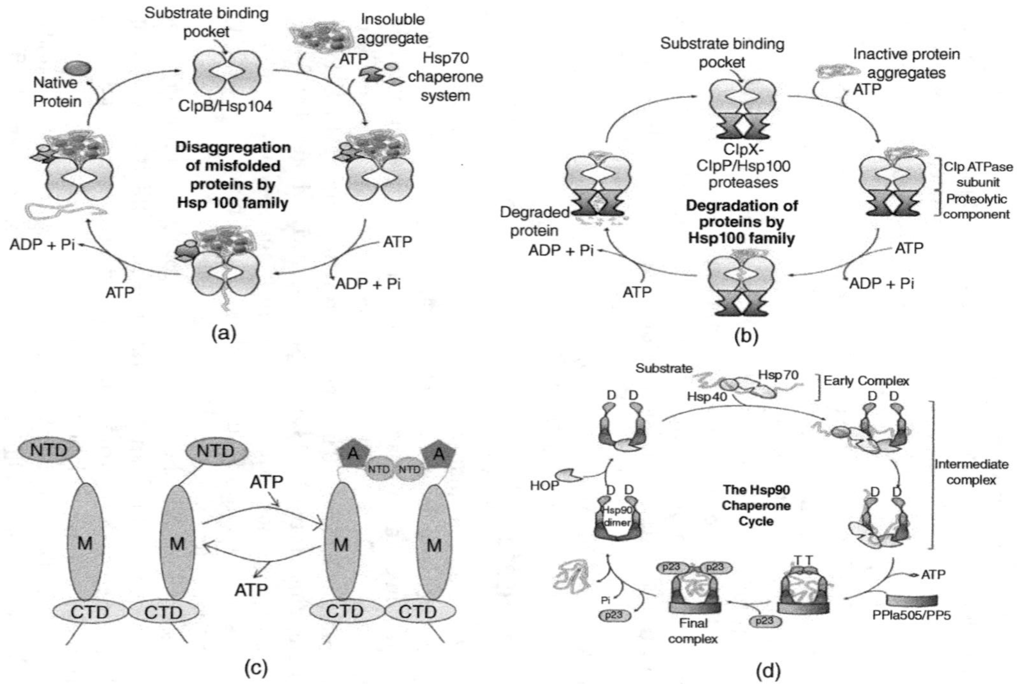

Figure 9.3 (a) HSP100 mechanism in protein disaggregation and (b) in protein degradation (http://pdslab.biochem.iisc.ernet.in/hspir/hsp100.php); (c) conformational change in HSP90; (d) HSP90 Mechanism (http://pdslab.biochem.iisc.ernet.in/hspir/hsp90.php).

of heat-induced genes, whereas in stress conditions HSP90.2 activates and induces the transcription of heat-induced genes.

9.2.2.1 Structure of HSP90

The Hsp90 family is highly conserved and found in eukaryotic cytosol. Cytosolic Hsp90s are further classified as Hsp90α and Hsp90β in humans (Xu et al. 2012), Hsc82 and Hsp82 in yeast, Hsp86 and Hsp84 in mice, and Hsp83 in *Drosophila* (Young et al. 2001). Hsp90αs involves in protein folding, protein transportation, maintenance of the normal conformation of protein, and preventing aggregation with ATP (Xu et al. 2012). Hsp90β is involved in the utilization of steroid hormones (Ogiso et al. 2004).

Hsp90 is characterized by structural domains: (i) the N-terminal domain (NTD), which contains the ATP-binding site and is important for dimerization; (ii) the middle domain (M), which is involved in the hydrolysis of ATP; (iii) the C-terminal domain, which is responsible for Hsp90 dimerization; and (iv) the C-terminal unstructured region, i.e. at the end of domain, has highly conserved Met-Glu-Glu-Val-Asp motif, and is involved in the co-chaperone recognition (Li and Buchner 2013; Jackson 2013) (Figure 9.2b).

9.2.2.2 Mode of Action: The HSP90 Chaperone Cycle

ATP binds to the NTD, which induces conformational changes in the HSP90s by hydrolysis of ATP which induces signal transduction, protein folding, and morphological evolution. Conformational changes in HSP90s occurs in such a way that N-domain open conformation changes to closed conformation with C-terminal domain (CTD) (Figure 9.3c). The CTD of Hsp90 is necessary for dimerization and continued rearrangements of the closed conformation allow interaction of the N-terminal and middle domains resulting in the "closed and twisted" conformation which is able to hydrolyze ATP into ADP (Xu et al. 2012). CTD has five peptide domains (MEEVD), which are necessary for interaction between HSP90 and proteins containing tetra trico peptide repeat (TPR) domains. Misfolded proteins or substrate binds to the Middle domain (M domain) because it has a substrate (peptide) binding region. After that P23, a small acidic protein with chaperone activity, combines with the middle section of the NTD of Hsp90 (Xu et al. 2012) and specifically identifies the binding state of Hsp90-ATP and induces the release of substrate protein in its native form (Figure 9.3d).

9.2.3 HSP70

The family of HSP acts as molecular chaperons and found in almost all plant and animal species. Hsp70 was originally discovered by F. M. Ritossa in the 1960s when the incubation temperature of *Drosophila* (fruit flies) was accidentally boosted (Ritossa 1996). When examining the chromosomes, Ritossa found a "puffing pattern" (uncoiled DNA) that indicated the elevated gene transcription for an unknown protein known as "heat shock proteins" (HSPs) (Ritossa 1996). The main function of HSP70s are (i) preventing aggregation of proteins and helping again in folding of proteins, (ii) stabilizing the newly formed proteins and preventing any possible misfolding and aggregation, (iii) importing proteins into cellular compartments such as mitochondria and endoplasmic reticulum.

9.2.3.1 Structure of HSP70

Hsp70 family protein contains two major domains; N-terminal ATPase domain (45 kDa) and substrate binding domain (SBD 18kD). N-terminal ATPase domain has a conventional nucleotide-binding fold (actin fold) which has a role in hydrolysis of ATP to ADP (Lund 2001). The second SBD is composed of subdomain β sheet and 10 kDa C-terminal helical subdomains. The SBD is subdivided into substrate binding subdomain or pocket and a C-terminal subdomain (rich in alpha helix and acts as a lid for the substrate binding pocket). This substrate-binding subdomain has a high affinity for neutral and hydrophobic amino acid residues (Lund 2001) (Figure 9.2c).

9.2.3.2 Mode of Action: The Hsp70 Chaperone Cycle

The Hsp70 function cycle is based on phenomena of peptide binding and ATP hydrolysis. These two activities occur in two different domains: the ATPase activity to the NBD and substrate binding to the CTD. In the Hsp70 chaperone cycle, substrate-binding occurs in a substrate binding pocket or subdomain of SBD and kinetics dependent upon the nucleotide state of the NBD where ATP bind. The ATP-bound form of Hsp70 has a low substrate binding affinity and a fast exchange rate, while the ADP-bound form exhibits higher affinity and slower exchange rates (Sharma and Masison 2009). The difference between the ATP- and ADP-bound states is the location of the CTD or lid, which covers the substrate-binding pocket when ADP is bound and thus reduces the release of bound substrate or peptides (Han and Christen 2003).

When ATP binds to the NBD it hydrolyzes into ADP and induces the conformational changes in the SBD and CTD, leading to substrate trapping in the substrate-binding subdomain or pocket. Correspondingly, substrate binding to the SBD alters the SBD structure in a way that transmits a signal to the NBD, which stimulates ATP hydrolysis, and J-domain protein also hydrolyzes ATP and thus increases its affinity of binding to Hsp70 (Sharma and Masison 2009). Subsequently, nucleotide exchange factors (NEFs) promote the release of ADP, facilitating again binding of ATP and thereby restoring Hsp70 to the low affinity state allowing substrate release. Released substrate is then refold in its native conformation or it binds again to Hsp70 and the remodeling cycle is repeated (Figure 9.4a).

9.2.4 HSP60

HSP60 is an ATP-dependent mitochondrial chaperone involved in the import of mitochondrial protein and macromolecular assembly. Mammalian HSP60 was first cloned and reported as a mitochondrial P1 protein (Gupta 1995). It promotes the correct folding of imported proteins. Under stress conditions, it prevents the misfolding of proteins and promotes the refolding and proper assembly of unfolded polypeptides in mitochondrial matrix. HSP60 amino acid sequence shows its similarity with GroEL (bacterial Hsp60 chaperone). Mitochondrial HSP60 contains a repeats of guanine (G) at the C-terminal and also has hydroxylated amino acids, namely serine, threonine, arginine, and lysine. HSP60 is classified into two groups (I and II) which are structurally similar but differ in sequences. Group I HSP60 is found in prokaryotes and endosymbiotic organelles such as mitochondria and chloroplasts of *Escherichia coli*. Group I HSP60 consists of two distinct subfamily proteins, chaperonin

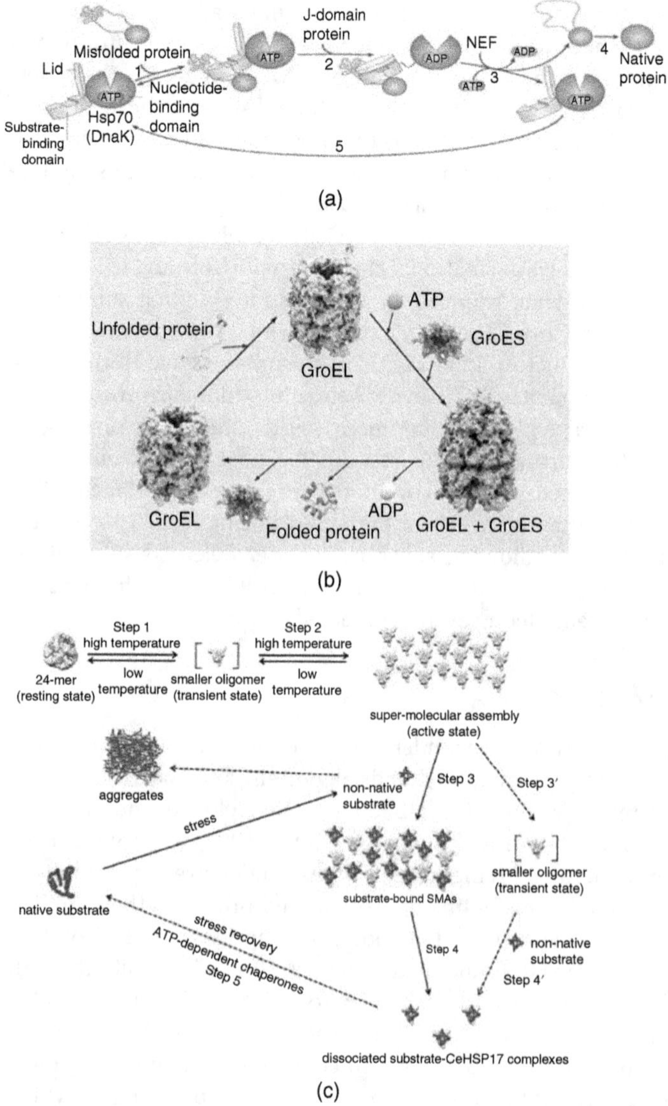

Figure 9.4 (a) HSP70 mechanism (Doyle et al. 2013); (b) HSP60 chaperone mechanism (Zhang et al. 2015); (c) Small HSP mechanism (Zhang et al. 2015).

60 (Cpn60) and chaperonin 10 (Cpn10) as co-chaperone, which function together in an ATP-dependent manner. Group II chaperonins exist in archaea and eukaryotic cytosol (Lund 2001) and enhance salt- and osmotic-stress tolerance of *E. coli* transformants (Yamada et al. 2002).

Plant chloroplast chaperonins have two distinct polypeptides, Cpn60a and Cpn60b, which are 50% similar in amino acid sequence. Cpn60a (chaperone 60a) and antisense Cpn60b influence chloroplast and plant embryo development, slow growth, delayed flowering, stunting and leaf chlorosis (Hemmingsen and Ellis 1986; Martel et al. 1990; Musgrove et al. 1987; Nishio et al. 1999).

9.2.4.1 Structure of HSP60

HSP60 forms two stacked heptameric rings with a central cavity to which substrate or unfolded protein binds through hydrophobic interactions. Each subunit of HSP60 has three domains: (i) the apical domain, (ii) the equatorial domain, and (iii) the intermediate domain (Figure 9.2d) (Ranford et al. 2000). HSP60 is based on these domains and cochaperone, for example GroES (sHSP). The equatorial domain contains ATP-binding site, whereas the intermediate domain connects the equatorial domain and the apical domain together (Bukau and Horwich 1998).

9.2.4.2 Mode of Action: The Hsp60 Chaperone Cycle

ATP bound to the equatorial domain which lead to conformational changes in the intermediate domain and form a cavity having a nucleotide-binding site which allows interaction between the substrate and substrate-binding sites. ATP bound to the equatorial domain which lead to conformational changes in the intermediate domain and form a cavity which has hydrophilic and hydrophobic substrate-binding site and nucleotide-binding site which allows interaction between the substrate and substrate-binding sites. ATP binding to the equatorial domain allows for an alternation between the hydrophilic and hydrophobic substrate-binding sites (Ranford et al. 2000). ATP binding also leads to conformational change in the intermediate domain which leads to exposure of the hydrophilic region. Chaperonin 10 (GroES) aids HSP60 (GroEL) in folding by acting as a dome-like cover on the ATP active form of HSP60, and when ATP hydrolyses into ADP, it releases Chaperonin 10 (GroES) and the folded protein (Figure 9.4b).

9.2.5 The Small Heat Shock Protein Family (sHSPs)

The sHSPs are an ancient and large family of proteins. They have been found in all domains of life such as Achaea, Eukarya, and Bacteria (Waters 2012). The sHSPs (15–42 kDa) form large oligomers having conserved sequence at their C-terminal. They are conserved across species and are important in stress tolerance. They work as a first line of defense against proteins misfolding and also act as co-chaperones for other higher HSPs, such as HSP70/HSP100, etc. The sHSPs have specific features that they do not require ATP to bind substrate proteins and they have very high capacity for binding with denatured proteins.

The number of sHSPs varies to individual plant species as *Arabidopsis* possesses 19, rice (*O. sativa*), 23 and poplar (*Populus trichocarpa*), and 36 sHSPs (Waters et al. 2008). The sHSPs cannot refold non-native proteins, but they can bind to partially folded or denatured substrates proteins, preventing irreversible unfolding or wrong protein aggregation (Mogk et al. 2003). Plants synthesize sHSPs when exposed to stresses such as HT, oxidative stress, drought, salts, etc. Therefore, sHSPs play an important role in stress management and protect the plant.

Small HSPs are divided into six classes based on their sequence alignments, cellular compartmentalization, and immunological cross reactivity. Three classes (I, II, and III) are localized in the cytoplasm or the nucleus and the remaining three are localized in the chloroplast, the endoplasmic reticulum, and the mitochondria (Sun et al. 2002; Vierling 1991).

9.2.5.1 Structure of sHSP

The structures of the sHSP family have been determined by X-ray crystallography or electron microscopy. The high-resolution crystal structures of Hsp16.5 from *Methanocaldo coccus* (*Methanococcus*) *jannaschii* Hsp16.9 from wheat (*Triticum aestivum*) were decoded by Kim et al. (2011) and Van et al. (2001), respectively.

The primary structure of sHSPs includes a variable N-terminal region, a more conserved C-terminal region, and a C-terminal extension. It has been suggested that the N-terminal region is important in substrate binding (Jaya et al. 2009; McDonald et al. 2012). The highly conserved C-terminal, also called α-crystallin domain (ACD), or HSP20 domain, contains organelle-specific retention amino acid motifs (Waters 2012) and plays a role in stabilizing the oligomer structure (Figure 9.2e). Crystal structure of sHsps is composed of two layers of three- and five-antiparallel strands, connected by a short interdomain loop.

9.2.5.2 Mode of Action: Small Heat Shock Proteins

sHSPs bind to the denatured proteins and prevent their irreversible aggregation (Jakob et al. 1993). Misfolded or unfolded proteins and sHSPs interact with each other, forming substrate–chaperone complexes. Small HSP has hydrophobic N-terminal regions which integrate and form oligomeric assembly and contribute to substrate binding and chaperone activity (Basha et al. 2012). The CeHSP17 (an sHSP from *Caenorhabditis elegans*) protein exists as spherical 24-mer (in resting state or inactive state) which lacks the chaperone-like activity at low temperatures. However, CeHSP17 (spherical 24-mer) transforms into the transient (unstable state) smaller oligomers when exposed to the HT, and unstable smaller oligomers may immediately self-associate into super-molecular assemblies (SMAs). N-terminal segments of SMAs interact with partially unfolded substrate proteins and thus prevent their aggregation or misfold (Zhang et al. 2015).

As the substrate binds to the N-domain it possibly triggers the dissociation of the small CeHSP17-substrate complexes from the substrate-bound SMAs, and the substrates in the small complexes are released in their native form with the help of ATP-dependent molecular chaperones like HSP70, 100, 104, etc. Alternatively, SMAs might function through the transient smaller oligomers (Zhang et al. 2015) (Figure 9.4c).

9.3 HSPs Expression Under Heat Stress Condition

Heat shock, stress is responsible for synthesis of HSPs. HSP mostly studied at approximately 10–20 °C higher than standard growth temperatures. Higher temperatures affect the grain yield by affecting the phenological development processes. Heat-induced yield reduction was well documented in several cultivated crops including cereals (e.g. rice, wheat, barley, sorghum, maize), pulses (e.g. chickpea, cowpea etc.), and oil yielding crops (mustard, canola), etc.

The rise of ambient temperatures due to global warming is a major cause of concern for heat stress to all crops (Grover et al. 2013 and Lavania et al. 2015). Heat stress affects the physiology of plants in various ways, including major reduction of net photosynthetic assimilation rate and changes in membrane characteristics (Avola et al. 2008; Hamada 2001). It also affects the pollen viability, reproductive stage, growth, and production in diverse legume and nonlegume crops.

HSP expression occurs by the presence of conserved heat shock elements (HSEs) in promoter region of HSGs (heat shock genes), which triggers transcription of HSGs under heat stress conditions. These HSEs have palindromic nucleotide sequence (5′-AGAANNTTCT-3′) that serve as a binding site for HSFs (Nover et al. 2001). It is observed that in most of the plant species HSFs are continuously expressed in monomeric form in the cytoplasm under normal conditions. Whenever plants have sensed heat stress (increase in temperature), the HSF monomer moves from the cytoplasm to the nucleus and forms a trimer that can bind with the HSEs and recruit the other transcriptional components, resulting in expression of HSP (Lee et al. 1995).

Studies have shown that heat-sensitive plants can be transformed to being heat resistant by gene transfer using genetic engineering techniques. Sarkar et al. (2014) show that the maintenance of homeostasis of cellular proteins is a critical component under the heat stress response (HSR) in rice seedlings. These heat shock proteins (Hsps) play a major role in protection of rice seedlings against heat stress.

It also has been observed that the expression of HSP101 in cotton plants under heat stress condition was responsible for elevated cotton boll production, growth, and pollen tube length (Burke and Chen 2015). AtHSP101 transgenic lines produced an average of 36 cotton bolls per plant, while the AtHSP101 negative null lines (absence of HSP101) produced only 12 bolls per plant (Burke and Chen 2015). Similarly, the numbers of seeds per plant produced by transgenic plants were also found significantly higher than those of null plants in elevated temperature environment. Thus the presence of HSP under heat stress condition maintains plant viability, lipid membranes, cytoskeletal components, nucleic acids in cells, proteins folding, and productivity.

9.4 Conclusion and Future Prospects

Plants are sessile organisms which cannot move to more favorable environments. Therefore, plant growth and developmental process are substantially affected, often lethally by HT stress. However, some plants species – such as cotton, chickpea, rice, wheat, corn, etc. – have the ability to respond to this stress by active utilization of HSPs, which relates to appropriate protein folding, prevention from disaggregation, signal transduction, transforming the misfolded protein into native protein, maintenance of plant growth and productivity, as well as thermotolerant capability of plants.

Heat stress disturbs cellular homeostasis and can lead to severe retardation of growth and development, and even death. Temperature stress can also have a devastating effect on plant metabolism. Therefore, understanding the molecular mechanism of HSPs/chaperones as regulatory molecules and its participation in stress sensing, signal transduction, and transcription activation of stress-related genes is very important in heat stress management and increasing crop productivity. Therefore, more research is needed to explore more opportunities to better understand plants' response to heat stress so that thermotolerant capability can be developed in heat-sensitive crop plants.

References

Avola, G., Cavallaro, V., Patane, C., and Riggi, E. (2008). Gas exchange and photosynthetic water use efficiency in response to light, CO_2 concentration and temperature in *Vicia faba*. *Journal of Plant Physiology* 165: 796–804. https://doi.org/10.1016/j.jplph.2007.09.004.

Banti, V., Maffessoni, F., Loreti, E. et al. (2010). Heat inducible transcription factor HsfA2 enhances anoxia tolerance in *Arabidopsis*. *Plant Physiology* 152: 1471–1483. https://doi.org/10.1104/pp.109.149815.

Basha, E., Lee, G.J., Breci, L.A. et al. (2004). The identity of proteins associated with a small heat shock protein during heat stress *in vivo* indicates that these chaperones protect a wide range of cellular functions. *Journal of Biological Chemistry* 27 (279): 7566–7575. https://doi.org/10.1074/jbc.M310684200.

Basha, E., O'Neill, H., and Vierling, E. (2012). Small heat shock proteins and α-crystallins, dynamic proteins with flexible functions. *Trends in Biochemical Sciences* 37: 106–117. https://doi.org/10.1016/j.tibs.2011.11.005.

Boston, R.S., Viitanen, P.V., and Vierling, E. (1996). Molecular chaperones and protein folding in plants. *Plant Molecular Biology* 32: 191–222. https://doi.org/10.1007/978-94-009-0353-1_9.

Bukau, B. and Horwich, L.A. (1998). The Hsp70 and Hsp60 chaperone machines. *Cell Press* 92: 351–366. https://doi.org/10.1016/S0092-8674(00)80928-9.

Bukau, B., Weissman, J., and Horwich, A. (2006). Molecular chaperones and protein quality control. *Cell* 125: 443–451. https://doi.org/10.1016/j.cell.2006.04.014.

Burke, J.J. and Chen, J. (2015). Enhancement of reproductive heat tolerance in plants. *PLoS One* 10: 1–23. https://doi.org/10.1371/journal.pone.0122933.

Campbell, J.L., Klueva, N.Y., Zheng, H.G. et al. (2001). Cloning of new members of heat shock protein HSP101 gene family in wheat (*Triticum aestivum* [L.] Moench) inducible by heat, dehydration, and ABA. *Biochimica et Biophysica Acta* (2): 270–277. https://doi.org/10.1016/S0167-4781(00)00292-X.

Chinnusamy, V. and Zhu, J.K. (2009). Epigenetic regulation of stress responses in plants. *Current Opinion in Plant Biology* 12: 133–139. https://doi.org/10.1016/j.pbi.2008.12.006.

Doyle, S.M. and Wickner, S. (2009). Hsp104 and ClpB, protein disaggregating machines. *Trends in Biochemical Sciences* 34: 40–48. https://doi.org/10.1016/j.tibs.2008.09.010.

Doyle, S.M., Genest, O., and Wickner, S. (2013). Protein rescue from aggregates by powerful molecular chaperone machines. *Nature Reviews Molecular Cell Biology* 14: 617–629. https://doi.org/10.1038/nrm3660.

Feder, M.E. and Hofmann, G.E. (1999). Heat-shock proteins, molecular chaperons, and the stress response, evolutionary and ecological physiology. *Annual Review of Physiology* 61: 243–282. https://doi.org/10.1146/annurev.physiol.61.1.243.

Fender, S.E. and Connell, M.A. (1990). Expression of the heat shock response in a tomato interspecific hybrid is not intermediate between the two parental responses. *Plant Physiology* 93: 1140–1146. https://doi.org/10.1104/pp.93.3.1140.

Glover, J.R. and Lindquist, S. (1998). Hsp104, Hsp70, and Hsp40, a novel chaperone system that rescues previously aggregated proteins. *Cell* 94: 73–82. https://doi.org/10.1016/S0092-8674(00)81223-4.

Gottesman, S., Squirest, C., Pichersky, E. et al. (1990). Conservation of the regulatory subunit for the Clp ATP-dependent protease in prokaryotes and eukaryotes. *Proceedings of the National Academy of Sciences of the United States of America* 87: 3513–3517.

Grover, A., Mittal, D., Negi, M., and Lavania, D. (2013). Generating high temperature tolerant transgenic plants, achievements and challenges. *Plant Science* 206: 38–47. https://doi.org/10.1016/j.plantsci.2013.01.005.

Gupta, R.S. (1995). Evolution of the chaperonin families (Hsp60, Hsp10 and Tcp-1) of proteins and the origin of eukaryotic cells. *Molecular Microbiology* 15: 1365–2958. https://doi.org/10.1111/j.1365-2958.1995.tb02216-X.

Hamada, A.M. (2001). Alteration in growth and some relevant metabolic processes of broad bean plants during extreme temperatures exposure. *Acta Physiologiae Plantarum* 23: 193–200.

Han, W. and Christen, P. (2003). Interdomain communication in the molecular chaperone DnaK. *Biochemical Journal* 369: 627–634. https://doi.org/10.1042/bj20020943.

Hemmingsen, S.M. and Ellis, R.J. (1986). Purification and properties of ribulose bisphosphate carboxylase large subunit binding protein. *Plant Physiology* 80: 269–276.

Jackson, S.E. (2013). Hsp90, structure and function. *Topics in Current Chemistry* 328: 155–240. https://doi.org/10.1007/128_2012_356.

Jakob, U., Gaestel, M., Engel, K., and Buchner, J. (1993). Small heat shock proteins are molecular chaperons. *Journal of Biological Chemistry* 268: 1517–1520.

Jaya, N., Garcia, V., and Vierling, E. (2009). Substrate binding site flexibility of the small heat shock protein molecular chaperons. *Proceedings of the National Academy of Sciences of the United States of America* 106: 15604–15609. https://doi.org/10.1073/pnas.0902177106.

Kim, D.H., Xu, Z.Y., Na, Y.J. et al. (2011). Small heat shock protein Hsp17.8 functions as an AKR2A cofactor in the targeting of chloroplast outer membrane proteins in *Arabidopsis*. *Plant Physiology* 157: 132–146. https://doi.org/10.1104/pp.111.178681.

Kumar, R., Lavania, D., Singh, A. et al. (2015). Identification and characterization of a small heat shock protein 17.9-CII gene from faba bean (*Vicia faba* L.). *Acta Physiologiae Plantarum* 37: 190. https://doi.org/10.1007/s11738-015-1943-3.

Larkindale, J., Hall, J.D., Knight, R.M., and Vierling, E. (2005). Heat stress phenotypes of *Arabidopsis* mutants implicate multiple signaling pathways in the acquisition of thermotolerance. *Plant Physiology* 138: 882–897. https://doi.org/10.1104/pp.105.062257.

Lavania, D., Dhingra, A., Siddiqui, M.H. et al. (2015). Current status of the production of high temperature tolerant transgenic crops for cultivation in warmer climates. *Plant Physiology and Biochemistry* 86: 100–108. https://doi.org/10.1016/j.plaphy.2014.11.019.

Lee, J.H., Hubel, A., and Schoffl, F. (1995). Derepression of the activity of genetically engineered heat shock factor causes constitutive synthesis of heat shock proteins and increased thermotolerance in transgenic *Arabidopsis*. *The Plant Journal* 8: 603–612. https://doi.org/10.1046/j.1365-313X.1995.8040603.x.

Li, J. and Buchner, J. (2013). Structure, function and regulation of the Hsp90 machinery. *Biomedical Journal* 36: 106–117. https://doi.org/10.4103/2319-4170.113230.

Ludwig-Muller, J., Krishna, P., and Forreiter, C. (2000). A glucosinolate mutant of *Arabidopsis* is thermosensitive and defective in cytosolic Hsp90 expression after heat stress. *Plant Physiology* 123: 949–958. https://doi.org/10.1104/pp.123.3.949.

Lund, P.A. (2001). *Molecular Chaperons in the Cell*. Oxford: Oxford University Press.

Martel, R., Cloney, L.P., Pelcher, L.E., and Hemmingsen, S.M. (1990). Unique composition of plastid chaperonin-60, alpha and beta polypeptide-encoding genes are highly divergent. *Gene* 94: 181–187.

McDonald, E.T., Bortolu, M., Koteiche, H.A., and McHaourab, H.S. (2012). Sequence, structure, and dynamic determinants of Hsp27 (HspB1) equilibrium dissociation are encoded by the N-terminal domain. *The Biochemist* 51: 1257–1268. https://doi.org/10.1021/bi2017624.

Mogk, A., Schlieker, C., Friedrich, K.L. et al. (2003). Refolding of substrates bound to small Hsps relies on a disaggregation reaction mediated most efficiently by ClpB/DnaK. *Journal of Biological Chemistry* 278: 31033–31042. https://doi.org/10.1074/jbc.M303587200.

Molinier, J., Ries, G., Zipfel, C., and Hohn, B. (2006). Transgeneration memory of stress in plants. *Nature* 442: 1046–1049. https://doi.org/10.1038/nature05022.

Musgrove, J.E., Johnson, R.A., and Ellis, R.J. (1987). Dissociation of the ribulose bisphosphate-carboxylase large-subunit binding protein into dissimilar subunits. *European Journal of Biochemistry* 163: 529–534.

Nishio, K., Hirohashi, T., and Nakai, M. (1999). Chloroplast chaperonins, evidence for heterogeneous assembly of alpha and beta Cpn60 polypeptides into a chaperonin oligomer. *Biochemical and Biophysical Research Communications* 266: 584–587.

Nover, L., Bharti, K., Doring, P. et al. (2001). *Arabidopsis* and the heat stress transcription factor world: how many heat stress transcription factors do we need? *Cell Stress and Chaperones* 6: 177–189. https://doi.org/10.1379/1466-1268,0062.0.CO;2.

Ogiso, H., Kagi, N., Matsumoto, E. et al. (2004). Phosphorylation analysis of 90 kDa heat shock protein within the cytosolic aryl hydrocarbon receptor complex. *The Biochemist* 43: 15510–15519. https://doi.org/10.1021/bi048736m.

Parsell, D.A. and Lindquist, S. (1993). The function of heat-shock proteins in stress tolerance, degradation and reactivation of damaged proteins. *Annual Review of Genetics* 27: 437–496. https://doi.org/10.1146/annurev.ge.27.120193.002253.

Queitsch, C., Hong, S.W., Vierling, E., and Lindquist, S. (2000). Heat shock protein 101 plays a crucial role in thermotolerance in *Arabidopsis*. *The Plant Cell* 12: 479–492. https://doi.org/10.1105/tpc.12.4.479.

Ranford, J.C., Coates, A.R., and Henderson, B. (2000). Chaperonins are cell-signalling proteins, the unfolding biology of molecular chaperons. *Expert Reviews in Molecular Medicine* 2: 1–17. https://doi.org/10.1017/S1462399400002015.

Ritossa, F. (1996). Discovery of the heat shock response. *Cell Stress and Chaperones* 1: 97–98.

Sarkar, N.K., Kim, Y.K., and Grover, A. (2014). Coexpression network analysis associated with call of rice seedlings for encountering heat stress. *Plant Molecular Biology* 84: 125–143. https://doi.org/10.1007/s11103-013-0123-3.

Sharma, D. and Masison, D.C. (2009). Hsp70 structure, function, regulation and influence on yeast prions. *Protein and Peptide Letters* 16: 571–581.

Sun, W., Montagu, M., and Verbruggen, N. (2002). Small heat shock proteins and stress tolerance in plants. *Biochimica et Biophysica Acta* 1577: 1–9. https://doi.org/10.1016/S0167-4781(02)00417-7.

Taipale, M., Jarosz, D.F., and Lindquist, S. (2010). HSP90 at the hub of protein homeostasis, emerging mechanistic insights. *Nature Reviews Molecular Cell Biology*: 515–528. https://doi.org/10.1038/nrm2918.

Usman, M.G., Rafii, M.Y., Ismail, M.R. et al. (2014). Heat shock proteins: functions and response against heat stress in plants. *International Journal of Advanced Research and Technology* 3: 204–218. https://doi.org/10.1155/2014/308042.

Van, M., Basha, R.L., Friedrich, E. et al. (2001). Crystal structure and assembly of a eukaryotic small heat shock protein. *Nature Structural Biology* 8: 1025–1030. https://doi.org/10.1038/nsb722.

Vierling, E. (1991). The roles of heat shock proteins in plants. *Annual Review of Plant Biology* 42: 579–620.

Wang, D. and Luthe, D.S. (2003). Heat sensitivity in a bentgrass variant: failure to accumulate a chloroplast heat shock protein isoform implicated in heat tolerance. *Plant Physiology* 133: 319–327. https://doi.org/10.1016/j.tplants.2004.03.006.

Wang, W., Vinocur, B., Shoseyov, O., and Altman, A. (2004). Role of plant heat-shock proteins and molecular chaperons in the abiotic stress response. *Trends in Plant Science* 9: 244–252. https://doi.org/10.1016/j.tplants.2004.03.006.

Waters, R.E. (2012). The evolution, structure and expression of the plant sHSPs. *Journal of Experimental Botany*: 391–403. https://doi.org/10.1093/jxb/ers355.

Waters, R.E., Nguyen, L.S., Eskandar, R. et al. (2008). The recent evolution of a pseudogene: diversity and divergence of a mitochondria-localized small heat shock protein in *Arabidopsis thaliana*. *Genome* 51: 177–186.

Xu, Z.-S., Li, Z.-Y., Chen, Y. et al. (2012). Heat shock protein 90 in plants, molecular mechanisms and roles in stress responses. *International Journal of Molecular Sciences* 13: 15706–15723. https://doi.org/10.3390/ijms131215706.

Yamada, A., Sekiguchi, M., Mimura, T., and Ozeki, Y. (2002). The role of plant CCTa in salt- and osmotic stress tolerance. *Plant and Cell Physiology* 43: 1043–1048. https://doi.org/10.1093/pcp/pcf120.

Young, J.C., Moarefi, I., and Hartl, F.U. (2001). Hsp90, a specialized but essential protein-folding tool. *Journal of Cell Biology* 154: 267–273. https://doi.org/10.1083/jcb.200104079.

Zhang, K., Ezemaduka, N.A., Wang, Z. et al. (2015). A novel mechanism for small heat shock proteins to function as molecular chaperons. *Scientific Reports* 5: 8811. https://doi.org/10.1038/srep08811.

Zhang, Q. and Denlinger, D.L. (2010). Molecular characterization of heat shock protein 90, 70 and 70 cognate cDNAs and their expression patterns during thermal stress and pupal diapause in the corn earworm. *Journal of Insect Physiology* 56: 138–150. https://doi.org/10.1016/j.jinsphys.2009.09.013.

10

The Contribution of Phytohormones in Plant Thermotolerance

Sonal Mishra[1], Mansi Bhardwaj[2], Shakti Mehrotra[3], Aksar Ali Chowdhary[4], and Vikas Srivastava[4]

[1] *School of Biotechnology, University of Jammu, Jammu, India*
[2] *ICAR-National Institute for Plant Biotechnology, IARI Campus, New Delhi, India*
[3] *Department of Biotechnology, Institute of Engineering and Technology, Lucknow, India*
[4] *Department of Botany, Central University of Jammu, Samba, India*

10.1 Introduction

The encounter of plants to their natural environment entails several abiotic stresses that affect many parts of the plant. Among other abiotic stresses, the temperature has been observed as a vital player for plant phenological development (Bahuguna and Jagadish 2015). Both low (cold and chilling stress) and high (heat) temperature have noteworthy impacts on crop plants. Further, global warming processes have significantly increased the problems associated with food security by significant reduction in crop production (Lobell et al. 2011). Most of the economically and dietary crops are temperature sensitive, and developing crops that can withstand adverse temperatures is a prerequisite to ensure the global food prospects. The rise in the earth's average temperature has affected the productive potential of many crop systems, which enforces studies to answer how plants sense and respond to increasing temperature.

Normally plants survive and show considerable growth and development in "physiological optimum" temperature. Beyond

Heat Stress Tolerance in Plants: Physiological, Molecular and Genetic Perspectives, First Edition.
Edited by Shabir Hussain Wani and Vinay Kumar.

this, growth and development are compromised and lead to a condition known as "high temperature or heat stress." HT prominently retards growth and development of plants, even after death (Huq and Nakamoto 2016). In fact, HT is one of the major determinants for plant growth, survival, and productivity success in the field. The immediate response to this involves the synthesis of various heat shock proteins (HSPs), which act as chaperone to prevent and repair cell damage and maintain homeostasis (Vierling 1991). The physiological threshold for the irreversible damage of plants by extremely HT varies significantly among plants and their genotypes. The HT stress may damage plant systems in various ways that lead to the instability of proteins, enzymes, nucleic acids, biomembranes, and cytoskeletal structures (Asthir 2015a,b).

The plant resists the impact of heat stress by developing various defense responses to minimize the impact of heat-induced damage (Song et al. 2012; Hemantaranjan et al. 2014). The thermotolerance in plants is of two types: basal and acquired. The basal thermotolerance is the inherent tolerant capacity to heat stress, here plants tolerate elevated temperatures without any heat acclimation (Clarke et al. 2004; Halter et al. 2016). The acquired thermotolerance involves a plant's ability to acquire tolerance, which is induced by either exposure to short but sub-lethal HT or a gradual rise in temperature to lethality, a condition similar to natural conditions (Song et al. 2012). Here, prior acclimatization to mild heat stress makes plants ready for any further warmer conditions. This type of acquired tolerance to heat in plants maintains the balance between growth and defense, as plant growth and subsequent yield are compromised, owing to any energy spent in response to heat stress (Salomé 2017). The foremost biochemical impact that heat stress is due to reactive oxygen species (ROS) mediated oxidative damage. ROS are dual feature toxic or signaling molecules, the role of which primarily depends on their concentration (Mishra et al. 2017). Further, HT-induced ROS generation causes damage to a variety of plant metabolites including lipid, proteins, and nucleic acids. The exogenous applications of several compounds have shown to mitigate the heat-stress-induced damage in plants (Hasanuzzaman et al. 2013). Amongst them, the contributions of plant hormones are well known to mitigate the impact of various stresses including heat. Various success stories of abscisic acid (ABA), auxin, gibberellins, cytokinins, salicylic acid (SA), jasmonic acid, ethylene, and brassinosteroids demonstrated their involvement in plant heat stress management and subsequent thermotolerance. Considering their significance, the chapter has been designed and will cover the significance of many plant hormones to combat the very unpredictable natural constraint of plant growth and success in the field.

10.2 Protectants in Heat Stress Alleviation

The administration of thermoprotectants such as osmolytes, phytohormones, polyamines, signaling molecule, trace elements, and nutrients (Figure 10.1) has been shown to offer positive effect on heat tolerance (Hasanuzzaman et al. 2013; Asthir 2015b). The exogenous applications of these are advantageous for plant growth under HT. Moreover, transgenic plants with improved potential to produce and accumulate thermoprotectants endogenously exhibit thermotolerance (Kaushal et al. 2016).

10.2.1 Osmolytes

Plants under stress conditions synthesize and accumulate friendly compounds called osmolytes or osmoprotectants. These small, electrically neutral, and nontoxic molecules are highly soluble organic compounds that maintain osmotic balance, stabilize proteins, and membranes, under variety of stresses (Nahar et al. 2016). Accumulation of osmolytes such as proline, glycine betaine, and trehalose is illustrious adaptive mechanism in plants against abiotic stress, including heat tolerance (Asthir 2015b; Kosar et al. 2018). During heat stress, the tolerance can be achieved by exogenous applications of osmolytes (Luo et al. 2010; Rasheed et al. 2011; Li et al. 2014).

10.2.2 Nutrients

Heat stress influences plant growth and development by enhancing the lipid membrane fluidity and protein denaturation. The drastic effects of HT can be mitigated by judicious

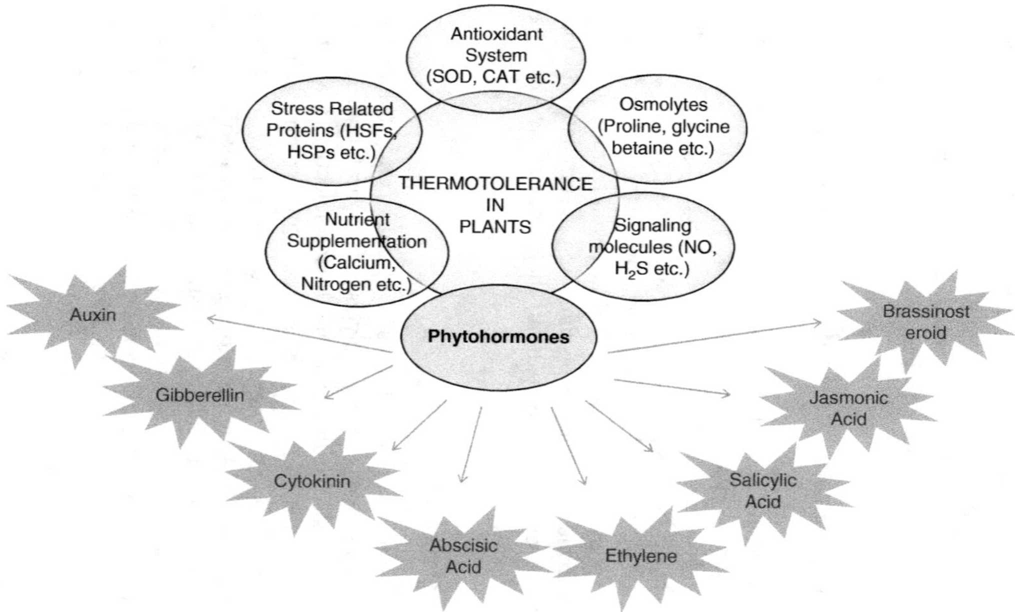

Figure 10.1 Strategies for heat stress mitigation in plants.

management of plant nutrition as most of the growth and development problems are associated with nutrient deficiency. Proper nutrient management also ensures the high growth and yield of the crop plants. Various nutrients – nitrogen (N), phosphorus (P), potassium (K), calcium (Ca), magnesium (Mg), selenium (Se), and zinc (Zn) – have been successfully applied to reduce the impact of temperature (Khalil et al. 2018). Under heat stress, the contribution of Ca in triggering the wheat thermotolerance through protein kinases-induced signal transduction is reported (Goswami et al. 2015). Balal et al. (2016) demonstrate the significance of Se in alleviation of heat-induced oxidative damage in cucumber, by improving growth parameters, antioxidant systems, osmolytes accumulation, nonphotochemical quenching, and lipid peroxidation.

10.2.3 Signaling Molecules

The signaling molecules (other than phytohormones) are also one of the imperative players during plant abiotic response and involved in stress responsive gene activation. These includes Ca-dependent protein kinases (CDPKs), mitogen-activated protein kinase (MAPK/MPKs), nitric oxide (NO), sugar as signaling molecule, etc. (Li et al. 2012; Parankusam et al. 2017). Song et al. (2006) show heat stress alleviation in *Phragmites communis* (reed) callus by two different NO donors, namely sodium nitroprusside (SNP) and S-nitroso-N-acetylpenicillamine (SNAP). The NO donor mitigates HT through alleviation of heat-stress-induced electrolyte leakage, growth retardation, and cell viability as well as H_2O_2 and MDA (malondialdehyde) contents. In this activity, NO donors activate the antioxidant system and this function is protected by NO scavenger (cPTIO). Pretreatment of sodium hydrosulfide (H_2S donor) demonstrated heat tolerance in tobacco suspension cells, which requires Ca^{2+} and calmodulin (Li et al. 2012).

10.2.4 Polyamines

Polyamines (PAs) are small, positively charged, ubiquitous organic molecules found in all organisms. In plants, PAs are represented by putrescine, spermidine, and spermine and in some plants by thermospermine (Minocha et al. 2014). These are widely identified in aspects related to stress tolerance in plants (Shi and Chan 2014; Sequera-Mutiozabal et al. 2017). Asthir and Deep (2011) report that putrescine pretreatment ameliorate the HT response in wheat and involves higher activities of antioxidant system. The application of spermine guards *Arabidopsis* plants from HT-induced damage, which was also observed by the application of other PAs (spermidine and thermospermidine) but not by putrescine (Sagor et al. 2013). The overexpressed lines with spermine synthase gene (SPMS) and its mutant *spms* revealed that higher spermine content offers higher thermotolerance.

10.2.5 Phytohormones

The application of hormone has been recognized as one of most imperative approaches for the amelioration of detrimental effects in heat-stressed plants (Ahammed et al. 2016). Larkindale and Knight (2002) observe that the exogenous application of ethylene precursor 1-aminocyclopropane-1-carboxylic acid (ACC), salicylic acid, and ABA offer protection to *Arabidopsis* from heat-induced oxidative damage. In their study, they observed increased susceptibility to heat stress by *etr-1* (ethylene-insensitive mutant), *abi-1* (ABA-insensitive mutant), and the transgenic line expressing *NahG* (an inhibitor of SA production). Several articles have undoubtedly emphasized the contribution of phytohormones in offering thermotolerance (Ahammed et al. 2016; Nazar et al. 2017).

10.3 Application of Hormones in HT Management

Plant hormones play an important role in plant growth and development. Apart from this, the endogenous content of many phytohormones has been found to change during adverse conditions, which offers clues as to the role phytohormones play as "stress responsive alleviators." Exogenous application of phytohormones has also demonstrated their contribution in heat stress alleviation (Figure 10.1). Further, contribution of phytohormones is also supported by heat sensitivity in mutants for phytohormone biosynthesis and signaling (Larkindale and Knight 2002; Clarke et al. 2004). The next section discusses the contribution of plant hormones in HT management.

10.3.1 Auxin

Auxin is a phytohormone which regulates cell division and growth in plants (Skoog and Miller 1957). Auxin, known to regulate diverse developmental processes, has also found having implications in abiotic and biotic stresses (Kazan 2013; Ahammed et al. 2016). Evidence for involvement of auxin in HT responses came from one of the pioneer studies by Gray et al. (1998). *Arabidopsis thaliana* seedlings grown at HT exhibited elongated hypocotyls with an increase in the concentration of auxin, suggesting temperature-mediated developmental responses require auxin. HT has been shown to elevate endogenous auxin in tobacco and cotton as well (Dobra et al. 2010; Min et al. 2014).

Conversely, reduced auxin was observed during HT stress in cultured guard cell protoplast (GCP) of *Nicotiana glauca* (Dong et al. 2007). At HT the GCP failed to regenerate walls as well as re-enter into the cell cycle, and suppressed induction of auxin-regulated genes. Heat stress causes suppression of auxin in developing anthers, resulting in abnormal pollens (Ozga et al. 2016). Auxin analogues reduce HT-induced lipid peroxidation, alleviating damage to membrane integrity in garden peas (Sergiev et al. 2018). Pretreatment of auxin analogues also keeps the level of H_2O_2 higher in heat-shocked plants as compared to the levels of H_2O_2 in untreated plants in order to maintain signaling cascade. Pretreatment of auxin analogues also keeps the level of antioxidant enzymes like superoxide dismutase (SOD), catalase (CAT), and peroxidase (POD) close to control upon HT (Sergiev et al. 2018). Also, proline content was higher in auxin-pretreated plants than that of HT-treated plants. These results show that auxin analogues help plants better adapt to HT stress in garden peas (Sergiev et al. 2018). Rice is highly sensitive to HT (Yoshida et al. 1981). HT stress results in reduced pollen viability, spikelet fertility, and yield (Yoshida et al. 1981). Endogenous auxin was found to be reduced upon HT stress

(Sharma et al. 2018). Exogenous application of indole-3-acetic acid (10^{-5} M) during heat/drought stress alleviated HT-induced lipid peroxidation of membrane, pollen viability, spikelet fertility, and yield (Sharma et al. 2018). Auxin biosynthesis gene YUCCA and its transcriptional regulators SQUAMOSA promoter-binding protein-like (SPL), NGATHA (NGA) and TERMINAL FLOWER (TFL) were downregulated in spikelets during HT stress (Sharma et al. 2018). Also, auxin receptors TRANSPORT INHIBITOR1 (TIR1)/AFB2-AUX/IAA, AUXIN RESPONSE FACTOR (ARF) target gene HAF, 3 PIN genes, and 5 ARFs were downregulated during stress in spikelets. HT not only reduced auxin biosynthesis but signaling and its transport as well as resulting in low endogenous auxin in spikelets (Sharma et al. 2018). The reduction in spikelet fertility under HT is because of inhibition of pollen tube elongation, owing to decrease in endogenous auxin in pistil (Zhang et al. 2018). The auxin level in pistil increases during fertilization implicating its role in pollen germination and elongation (Wu et al. 2008). To establish the relationship between auxin and pollen germination, heat stress was given to two genotypes of rice differing in heat tolerance, i.e. Nipponbare (NPB) and high temperature sensitive (HTS) (Zhang et al. 2018). Exogenous application of 1-naphthaleneacetic acid (NAA) led to reversal of HT-induced spikelet sterility in NPB but not in HTS. Also, exogenous NAA-induced endogenous IAA production and ROS in pistil of NPB under HT. The increase in endogenous IAA rescued spikelet sterility by preventing inhibition of pollen tube germination (Zhang et al. 2018). A decrease in levels of endogenous auxin at HT is due to inhibition of expression of auxin biosynthesis gene YUC2 and YUC6 (Chen et al. 2006; Sakata et al. 2010). The decrease in auxin causes abortion of pollen development, resulting in a decrease in yield. In wheat, barley, and *Arabidopsis*, HT causes male sterility by arresting cell cycle proliferation and disruption of auxin signaling specific to anther (Higashitani 2013). Exogenous application of auxin complements the normal pollen grains indicating auxin is one of the key cues in normal development (Higashitani 2013). Synthetic auxins can be used to rescue male fertility and yield under threat of global warming.

10.3.2 Gibberellin

One of the earlier studies showed that barley showed heat tolerance in the absence of gibberellins (Grindstaff et al. 1996). *Hordeum vulgare* aleurone endoplasmic reticulum (ER membrane) showed heat-tolerance in the absence of GA_3 (Grindstaff et al. 1996). GA_3-treated, heat shocked aleurone layer of barley resulted in suppression of secreted proteins. The suppression of secreted proteins was primarily due to a high degree of unsaturated fatty acid content in endoplasmic reticulum of the aleurone. Contrastingly, increase in saturation of aleurone ER membrane was observed in the absence of GA_3, which helps the aleurone layer sustain its function during heat shock. On the contrary, it was shown that GA_3 regulates thermotolerance and helps plants in combatting heat stress by inducing synthesis and action of salicylic acid (SA) in *A. thalian*. Exogenously applied GA_3 was able to reverse the effects of abiotic stress like salinity, oxidative stress, and heat stress in germination of *A. thaliana*. Also, GA_3 treatment enhanced expression of SA biosynthesis and action genes, i.e. isochorismate synthase (SA synthesis) and nonexpressor of PR1 (SA action). Thus GA induces SA production, which helps plants thrive during heat stress along with other abiotic stresses.

More evidence of phytohormones being involved in thermotolerance comes from the study of heat-sensitive and heat-tolerant varieties of rice by Wu et al. (2016). Four varieties of rice, namely Nagina22, Lianyoupeijiu, Huanghuazhan, and Shanyou63, were heat-treated, which led to reduced level of GAs, IAA, and cytokinins. Reduction in level of IAA and GAs resulted in reduced spikelet fertility, owing to reduction in pollen vigor and anther dehiscence (Wu et al. 2016). As a result of reduced cytokinins, IAA, and GA reduced grain size and grain weight were observed. The heat-resistant variety SY63 exhibited stable levels of IAA, GA1, tZ + tZR in young panicles, resulting in heat-resistant panicle (Wu et al. 2016). Thus phytohormone does play a role in protection of plants against heat stress.

Phytochrome interacting factor (PIF) acts as a node of phytohormone signaling by integrating GA, auxin signaling with PIF4 at HT. PIF4 is a key thermoregulator (Koini et al. 2009). GA regulates PIF4 via destabilization of DELLA proteins responsible for the inhibition of PIF4 (De Lucas et al. 2008; Feng et al. 2008). Plants thrive during stress by shortening their lifecycle, thereby inducing early flowering. Elevated temperatures also induce flowering. This temperature-induced early flowering depends upon flowering locus T (FT). During heat stress PIF4 activates FT bringing about flowering (Kumar et al. 2012). As inhibition of GA biosynthesis in *A. thaliana* resulted in alteration of early flowering (Galvão et al. 2015); therefore, GA may regulate flowering at HT via PIF4 activation of FT (Kumar et al. 2012).

10.3.3 Cytokinin

Cytokinin is a cell division and growth-regulating phytohormone (Skoog and Miller 1957). Cytokinin not only regulates growth but also responds to abiotic stresses. HT has become a serious concern, because of global warming. Heat stress is known to affect grain growth and yield stability to a significant extent in *Zea mays* (Cheikh and Jones 1994). The heat-stress-induced reduction in yield of maize is dissected by Cheikh and Jones (1994). ABA and cytokinin levels were found to be altered in heat-stressed kernels as compared to control. Cytokinins zeatin and zeatin riboside were found to be 70% reduced in heat-stressed kernels compared to control. Exogenous application of benzyl adenine to heat-stressed kernels showed thermotolerance by rescuing kernel abortion at tip and middle of ear. This study established the significance of cytokinin in thermotolerance and yield stability of maize (Cheikh and Jones 1994). Wheat is sensitive to temperature and requires optimum temperature for reproduction and grain production. HT leads to reduction in grain yield in wheat (Banowetz et al. 1999). A marked decline in cytokinin was observed in wheat kernel upon heat stress accompanied by reduction in kernel weight (Banowetz et al. 1999). Cytokinin level was studied in Tibet Dwarf (*Triticum aestivum*), known to accumulate higher levels of cytokinin in leaf. It was observed that cytokinin levels dropped to 80% in heat-stressed kernels compared to cytokinin level of untreated plants (Banowetz et al. 1999). The reduced cytokinin content of heat-treated kernels were still higher than cytokinin content of untreated plants of other varieties. Thus cytokinin alone is not responsible for the thermotolerance of kernels, but it is the combination of cytokinin and genotype that confers tolerance to elevated temperatures (Banowetz et al. 1999). Effect of HT on cytokinin level was studied on model plant *A. thaliana* (Todorova et al. 2005). It was observed that HT had a more pronounced effect on cytokinin level as compared to low temperature (Todorova et al. 2005). At HT, cytokinin was found to decrease, whereas, at low temperature, cytokinin levels were elevated in *A. thaliana* (Todorova et al. 2005). In wheat, HT not only reduces the number of kernels per spike but also reduces number of endosperm cells significantly ($P < 0.05$) (Yang et al. 2016). Stay green Wennong 6 and nonstay green variety Jimai 20 of wheat were analyzed under heat stress. It was found that endosperm of nonstay green variety was susceptible to heat stress unlike the stay green variety. (Yang et al. 2016) Heat stress causes reduction in endosperm cell division, owing to a decrease in zeatin and increased GA_3 levels which negatively affects grain filling in heat-susceptible variety. The stay green variety could provide nutrition to grain during heat stress due to high cytokinin and normal endosperm resulting in better yield. Exogenous application of 6-benzyl adenine induces cell division in endosperm thereby high grain filling alleviating effects of heat stress (Yang et al. 2016).

Molecular mechanism of reduced cytokinin upon heat stress in rice was studied by Wu et al. (2017) in three heat-susceptible varieties of rice, namely Nagina22, Lianyoupeijiu, Huanghuazhan, and one heat-tolerant variety Shanyou63. A significant reduction in the level of cytokinin was observed in heat-susceptible varieties of rice upon heat stress (Wu et al. 2017). This marked reduction in cytokinin was due to a reduced transport of cytokinin from root to shoot accompanied by enhanced activity of cytokinin degrading enzyme cytokinin oxidase (Wu et al. 2017). As a result of reduced cytokinin fewer spikelets per panicle were observed. Contrastingly, heat-tolerant variety SY63 exhibited comparatively less reduction in cytokinin levels of panicle, owing to better transport of root-derived cytokinin. Also, heat-tolerant variety SY63 showed stable panicle in heat-treated plants. Cytokinins are a potential target for developing heat-tolerant plants in rice (Wu et al. 2017).

10.3.4 Abscisic Acid

The ABA is a sesquiterpene stress hormone known to offer resistance to abiotic stresses, including heat. The production of ABA increases during heat stress treatment, which subsequently offers ABA-induced thermotolerance. While studying thermotolerance in maize seedlings, Gong et al. (1998) suggest that the ABA-induced thermotolerance involves Ca^{2+}, entry of extracellular Ca^{2+} into cells across the plasma membrane, and antioxidant enzyme systems. In this study, they observe that $CaCl_2$ enhances ABA-induced thermotolerance, whereas treatment with Ca^{2+} chelator EGTA decreases the ABA-induced thermotolerance. Besides, applications of Ca^{2+} channel blockers La^{3+} and verapamil also compromised the ABA-induced thermotolerance. During heat-stress-induced oxidative stress in maize seedling, the treatment of ABA was found to induce the activities of antioxidant enzymes (SOD, CAT, APX, and GPX) and diminishes the lipid peroxidation. These activities were further promoted by Ca^{2+} pretreatment. Larkindale and Knight (2002) have also demonstrated the role of ABA in the protection of *Arabidopsis* against heat-induced oxidative damage along with ethylene, and SA. Recent studies reveal that plants respond differently to individual environmental stress and the combination of such stresses. *Arabidopsis* was found to be more susceptible in a combination of salt and drought stress than their individual treatments. This combinatorial stress exhibited increased ratio of Na^+/K^+ along with enhanced accumulation of 699 transcripts, out of them many were associated with ABA. Further, ABA biosynthesis (*aba1-1*) and signaling deficient (*abi1-1*) mutant were more susceptible to this combination than wild plants (Suzuki et al. 2016). Li and Jin (2016) observe that ABA could improve many heat-stress-affected properties of tobacco suspension such as survival percentage and regrowth ability, and also mitigate cell vitality reduction, malondialdehyde content, and electrolyte leakage induction. The ABA treatment also enhanced the accumulation of endogenous H_2S by upregulating of L-cysteine desulfhydrase activity. Additionally, the ABA-induced heat tolerance was further improved by NaHS (H_2S donor) and weakened by H_2S biosynthesis inhibitor (DL-propargylglycine) and scavenger (hypotaurine), suggesting

ABA-induced thermotolerance is partly mediated by H_2S. The heat exposures at meiotic pollen mother cell stage lead to pollen sterility in rice, in which ROS and sugar homeostasis are negatively affected. This adverse effect (pollen sterility) can be improved by ABA through enhanced content of soluble sugars, starch, nonstructural carbohydrates and adenosine triphosphate, upregulation of HSPs, and genes associated with sugar metabolism and transport, enhanced antioxidant activities, and endogenous ABA in spikelets (Rezaul et al. 2019).

10.3.5 Ethylene

Ethylene is a gaseous phytohormone that is involved in many aspects of plant growth as well as in plant responses to biotic and abiotic stresses. Larkindale and Knight (2002) examine the protective effect of calcium, ABA, ethylene, and SA against heat-induced oxidative damage in *Arabidopsis*. Studies with mutants, namely *etr-1* (ethylene-insensitive), *abi-1* (ABA-insensitive mutant), and *NahG* (SA production inhibition) overexpressing transgenic lines demonstrated increased susceptibility to heat. The treatment of SA, H_2O_2, ACC (an ethylene precursor), or ABA can induce thermotolerance (Larkindale and Huang 2005). Further, increased production of SA and H_2O_2 was observed during heating initiation and ethylene and ABA during recovery period. It has been shown by Clark et al. (2009) that ethylene acts as an antagonistic to JA. In their investigation with *ein2* mutant (ethylene insensitive), they noticed thermotolerance, which suggested negative regulation of heat stress by EIN 2 mediated pathway.

10.3.6 Salicylic Acid

2-hydroxy benzoic acid (SA) is a small phenolic compound involved in plant growth, development, and attainment of stress tolerance to various abiotic stress. Its function has also been observed in stress related to temperature such as cold/chilling and also in heat stress (Nazar et al. 2017). Many plant species have been reported for enhance thermotolerance after supplementation of SA (Senaratna et al. 2000; Shi et al. 2006; Hameed and Ali 2016; Abd-Elkader et al. 2016;

Lin et al. 2019; Jahan et al. 2019). The improved plant heat-stress tolerance has been observed through mediation of SA in a number of plants (Larkindale and Huang 2005; Wang et al. 2010). The pretreatment of SA, alleviated heat-stress-induced decrease in net photosynthesis rate (Wang et al. 2010). Both the accumulation of SA and its signaling are important in basal thermotolerance. Senaratna et al. (2000) observed that seeds of bean and tomato imbibed in aqueous solutions (0.1–0.5 mM) of SA or acetyl salicylic acid (ASA) displayed enhanced tolerance to multiple stresses, including heat. In a study, Larkindale and Knight (2002) demonstrated that many hormones participate in heat-stress tolerance. The *NahG* (coding for a bacterial salicylate hydroxylase) expressing transgenic plants that did not permit SA accumulation was found sensitive to heat stress. Further, application of SA provides defense against heat-induced oxidative damage in wild types of *Arabidopsis* plants. The role of SA has also been observed while analyzing the thermotolerance of *Arabidopsis*. In this study, lowest thermotolerance was observed in SA-deficient *NahG* transgenic plants and highest thermotolerance was observed with SA-accumulating *cpr5* (*constitutive expressor of PR genes*) mutants. (Clarke et al. 2004, 2009). The *npr-1* (*nonexpressor of PR1 proteins*) and *npr-5* signaling mutants of *A. thaliana* are prone to heat stress (Ahammed et al. 2016). However, the *cpr-5* (*constitutive expressers of PR proteins*) mutant that was constitutively active for SA, JA, and ethylene-signaling pathways exhibited enhanced thermotolerance (Clarke et al. 2009).

Further, it has been observed that osmolytes (proline) make a significant contribution to heat stress alleviation. Salicylic has also been shown to induce proline production, which is also regulated by its interaction with ABA and ethylene (Iqbal et al. 2019). Feng et al. (2018) worked on heat stress on rice cultivar and found that SA could also mediate the alleviation of damage to rice pollen viability. During heat stress, SA reduces ROS accumulation in anthers and thus checks programmed cell death of tapetum. Further, higher pollen viability was observed with the supplementation of either SA or paclobutrazol (PAC, a SA inhibitor) + H_2O_2 under heat stress as compared to plants sprayed with either H_2O_2, PAC, or SA + dimethylthiourea (DMTU, an H_2O_2 and $OH^{\cdot}$

scavenger), which suggested contribution of H_2O_2 in mediating SA function to prevent pollen abortion through inhibiting the tapetum PCD during heat stress. Li et al. (2015) have observed improvement of maize seedling survival during heat stress by SA pre-treatment. Further, the supplementation of SA enhances endogenous H_2S content and activity by enhancing the activity of L-cysteine desulfhydrase (L-DES). SA-induced heat tolerance was further elevated by NaHS treatment, and weakened by H2S inhibitor (DL-propargylglycine) and scavenger (hydroxylamine). However, SA biosynthesis inhibitors (paclobutrazol and 2-aminoindan-2-phosphonic acid) treatment had no effect on NaHS-induced heat tolerance and no significant change in endogenous SA content or enzymes associated with SA biosynthesis were noted. These investigations suggested possible downstream activity of H_2S in SA-induced heat tolerance in maize seedlings. In a field study using wheat genotypes, Chouhan et al. (2017) demonstrate that SA treatment alleviates the HT effect on wheat yield and yield attributes. Kumar et al. (2015) observed that SA treatment could alleviate detrimental effect of HS on soluble starch synthase (SSS) activity and starch granules formation in wheat cultivar. The NAC transcription factor of tomato (SlJA2) was overexpressed in tobacco plants (Liu et al. 2017), which showed decrease in SA accumulation and heat tolerance, owing to changes in stress physiology and expression of genes. The study of heat stress on spikelet degeneration of rice was studied by Zhang et al. (2017). Under heat stress, SA treatment exhibited a higher grain yield, spikelet number per panicle and setting rate as compared to control. Further, the induction of soluble sugars, proline, phytohormones, and antioxidant enzymes was also observed during the process.

10.3.7 Jasmonic Acid

It is a class of lipid phytohormone involved in variety of plant functions particularly in development, stress response, and plant–microbe interaction. Many investigations have suggested the role of JA in heat-stress tolerance (Chen et al. 2006; Clark et al. 2009; Sharma and Laxmi 2016). Clark et al. (2009) in *Arabidopsis* show the positive role of JA in basal thermotolerance, which was evident in compromised heat tolerance in *cpr5-jar1* double mutant. However, the wild type plants are protected from heat stress by an exogenous application of methyl jasmonate.

10.3.8 Brassinosteroid

Brassinosteroids (BRs) are a novel group of phytohormones that includes polyhydroxy steroidal compounds with significant biological roles in increasing crop yields through mitigating various environmental stresses. One active BR is 24-epibrassinolide (EBL) that can stimulate many physiological and biochemical processes such as carbon assimilation, biosynthesis of nucleic acids, ROS metabolism, and light harvesting (Siddiqui et al. 2018). It also assists plants to mitigate various adverse conditions, including heat stress through the regulation of plant metabolism (Janeczko et al. 2011; Anjum et al. 2011). Dhaubhadel et al. (2002) observed that treatment of EBL improved the basic thermotolerance in *Brassica napus* seedlings through synthesis and accumulation of HSPs. During the course of thermal stress, BRs ensure HSPs biosynthesis, through offering protection to translational machinery. In heat-stressed tomato plants, the BRs (EBR and MH5) contributed to reduce cell damage, owing to induction antioxidants enzyme activity (Mazorra et al. 2002). Bajguz (2009) demonstrates that exogenously applied BRs can stimulate production of ABA in *Chlorella vulgaris* subjected to short-term heat stress. The exogenous application of BRs (EBL) alleviates HT-induced inhibition of photosynthesis (increase carboxylation efficiency), improved antioxidant system and reduced total hydrogen peroxide (H_2O_2) and malonaldehyde (MDA) contents. Additionally, BRs also help in the increase of shoot weight, thus also alleviating the detrimental impact of heat stress in tomato plants (Ogweno et al. 2008). While working on rice, Thussagunpanit et al. (2015a) report that 7,8-dihydro-8α-20-hydroxyecdysone (DHECD) – a BR mimic can alleviate heat stress similar to EBR by improving growth, photosynthesis, and total soluble sugar content and reducing malondialdehyde and hydrogen peroxide contents. Further, the application of 1 nM EBR and/or 1 nM DHECD at the reproductive stage during the hot season could potentially increase the rice yield (number of filled seeds) (Thussagunpanit et al. 2015b). In tall fescue (*Festuca arundinacea* Schreb.) the pretreatment with melatonin (MT) and 24-epibrassinolide (EBL) improved thermotolerance efficiency by reducing ROS, electrolyte leakage and MDA content, and increasing chlorophyll (Chl), total protein, and antioxidant enzyme activities (Alam et al. 2018).

10.4 Conclusion and Prospects

Global warming has become one of the major concerns of recent times. HT stress affects photosynthesis, reproduction, yield, etc. Phytohormones have provided an effective solution to manage plant survival under heat stress. Topical applications of phytohormones mitigate the effects of heat stress and reverse the HT-induced negative effects, thereby rescuing the plant. Further, a plant with inefficient hormone system (either synthesis or signaling) was found susceptible to heat stress. In this direction, contribution of various hormones has been established and observed to work in coordination with other signaling network while offering tolerance to heat. However, some hormones need further exploration in this direction like strigolactone, ethylene, and jasmonic acid, which either have not been explored or their exploration is limited to only a few studies. Further, it is now imperative to look into the order to participation of these hormones and to answer the questions: what regulates what and at which condition and how is the activity of one influenced by another? This kind of investigation will give insight to identify the master regulatory pathway that will be utilized to make more effective strategy for developing crops for efficient temperature management.

Acknowledgements

V.S. acknowledges Central University of Jammu for providing the work facility and also thankful to UGC, India for Start-Up research grant. S.M. is thankful to Science & Engineering Research Board (SERB, India) for NPDF (PDF/2016/002133) fellowship.

References

Abd-Elkader, A.M., Mahmoud, M.M., Shehata, S.A. et al. (2016). Induction of thermotolerant tomato plants using salicylic acid and kinetin foliar applications. *Journal of Horticultural Science & Ornamental Plants* 8 (2): 89–97.

Ahammed, G.J., Li, X., Zhou, J. et al. (2016). Role of hormones in plant adaptation to heat stress. In: *Plant Hormones under Challenging Environmental Factors* (eds. G.J. Ahammed and J.Q. Yu), 1–21. Dordrecht, the Netherlands: Springer.

Alam, M.N., Zhang, L., Yang, L. et al. (2018). Transcriptomic profiling of tall fescue in response to heat stress and improved thermotolerance by melatonin and 24-epibrassinolide. *BMC Genomics* 19 (1): 224.

Anjum, S.A., Wang, L.C., Farooq, M. et al. (2011). Brassinolide application improves the drought tolerance in maize through modulation of enzymatic antioxidants and leaf gas exchange. *Journal of Agronomy and Crop Science* 197 (3): 177–185.

Asthir, B. (2015a). Mechanisms of heat tolerance in crop plants. *Biologia Plantarum* 59 (4): 620–628.

Asthir, B. (2015b). Protective mechanisms of heat tolerance in crop plants. *Journal of Plant Interactions* 10 (1): 202–210.

Asthir, B. and Deep, A. (2011). Thermotolerance and antioxidant response induced by putrescine and heat acclimation in wheat seedlings. *Seed Science Biotechnology* 5 (1): 42–46.

Bahuguna, R.N. and Jagadish, K.S. (2015). Temperature regulation of plant phenological development. *Environmental and Experimental Botany* 111: 83–90.

Bajguz, A. (2009). Brassinosteroid enhanced the level of abscisic acid in *Chlorella vulgaris* subjected to short-term heat stress. *Journal of Plant Physiology* 166 (8): 882–886.

Balal, R.M., Shahid, M.A., Javaid, M.M. et al. (2016). The role of selenium in amelioration of heat-induced oxidative damage in cucumber under high temperature stress. *Acta Physiologiae Plantarum* 38 (6): 158.

Banowetz, G.M., Ammar, K., and Chen, D.D. (1999). Temperature effects on cytokinin accumulation and kernel mass in a dwarf wheat. *Annals of Botany* 83 (3): 303–307.

Cheikh, N. and Jones, R.J. (1994). Disruption of maize kernel growth and development by heat stress (role of cytokinin/abscisic acid balance). *Plant Physiology* 106 (1): 45–51.

Chen, P.Q., Yu, S.L., Zhan, Y.N., and Kang, X.L. (2006). Effects of jasmonate acid on thermotolerance of grape seedlings. *Journal of Shihezi University (Natural Science)* 24: 87–91.

Chouhan, K.S., Kakralya, B.L., Bajya, M., and Sodani, R. (2017). Salicylic acid mitigate the adverse effect of high temperature stress on yield and yield determining parameters of wheat (*Triticum aestivum* L.). *Journal of Pharmacognosy and Phytochemistry* 6 (4): 1052–1055.

Clarke, S.M., Mur, L.A., Wood, J.E., and Scott, I.M. (2004). Salicylic acid dependent signaling promotes basal thermotolerance but is not essential for acquired thermotolerance in *Arabidopsis thaliana*. *The Plant Journal* 38 (3): 432–447.

De Lucas, M., Daviere, J.M., Rodríguez-Falcón, M. et al. (2008). A molecular framework for light and gibberellin control of cell elongation. *Nature* 451 (7177): 480.

Dhaubhadel, S., Browning, K.S., Gallie, D.R., and Krishna, P. (2002). Brassinosteroid functions to protect the translational machinery and heat-shock protein synthesis following thermal stress. *The Plant Journal* 29 (6): 681–691.

Dobra, J., Motyka, V., Dobrev, P. et al. (2010). Comparison of hormonal responses to heat, drought and combined stress in tobacco plants with elevated proline content. *Journal of Plant Physiology* 167: 1360–1370.

Dong, M.A., Bufford, J.L., Oono, Y. et al. (2007). Heat suppresses activation of an auxin-responsive promoter in cultured guard cell protoplasts of tree tobacco. *Plant Physiology* 145 (2): 367–377.

Feng, S., Martinez, C., Gusmaroli, G. et al. (2008). Coordinated regulation of *Arabidopsis thaliana* development by light and gibberellins. *Nature* 451 (7177): 475.

Feng, B., Zhang, C., Chen, T. et al. (2018). Salicylic acid reverses pollen abortion of rice caused by heat stress. *BMC Plant Biology* 18 (1): 245.

Galvão, V.C., Collani, S., Horrer, D., and Schmid, M. (2015). Gibberellic acid signaling is required for ambient temperature-mediated induction of flowering in *Arabidopsis thaliana*. *The Plant Journal* 84 (5): 949–962.

Gong, M., Li, Y.J., and Chen, S. (1998). Abscisic acid-induced thermotolerance in maize seedlings is mediated by calcium and associated with antioxidant systems. *Journal of Plant Physiology* 153 (3–4): 488–496.

Goswami, S., Kumar, R.R., Sharma, S.K. et al. (2015). Calcium triggers protein kinases-induced signal transduction for augmenting the thermotolerance of developing wheat (*Triticum aestivum*) grain under the heat stress. *Journal of Plant Biochemistry and Biotechnology* 24 (4): 441–452.

Gray, W.M., Ostin, A., Sandberg, G. et al. (1998). High temperature promotes auxin-mediated hypocotyl elongation in *Arabidopsis. Proceedings of the National Academy of Sciences* 95: 7197–7202.

Grindstaff, K.K., Fielding, L.A., and Brodl, M.R. (1996). Effect of gibberellin and heat shock on the lipid composition of endoplasmic reticulum in barley aleurone layers. *Plant Physiology* 110 (2): 571–581.

Halter, G., Simonetti, N., Suguitan, C. et al. (2016). Patterns of thermotolerance, chlorophyll fluorescence, and heat shock gene expression vary among four *Boechera* species and *Arabidopsis thaliana. Botany* 95 (1): 9–27.

Hameed, S. and Ali, M.K. (2016). Exogenous application of salicylic acid: inducing thermotolerance in cotton (*Gossypium hirsutum* L.) seedlings. *International Journal of Agricultural and Food Research* 5 (4).

Hasanuzzaman, M., Nahar, K., Alam, M.M. et al. (2013). Physiological, biochemical, and molecular mechanisms of heat stress tolerance in plants. *International Journal of Molecular Sciences* 14 (5): 9643–9684.

Hemantaranjan, A., Bhanu, A.N., Singh, M.N. et al. (2014). Heat stress responses and thermotolerance. *Advances in Plants & Agriculture Research* 1 (12): 10–15406.

Higashitani, A. (2013). High temperature injury and auxin biosynthesis in microsporogenesis. *Frontiers in Plant Science* 4: 47.

Huq, S. and Nakamoto, H. (2016). Heat shock proteins and acquisition of thermotolerance in plants. In: *Handbook of Plant and Crop Stress* (ed. M. Pessarakli), 529–544. CRC Press.

Iqbal, N., Fatma, M., Khan, N.A., and Umar, S. (2019). Regulatory role of proline in heat stress tolerance: modulation by salicylic acid. In: *Plant Signaling Molecules* (eds. M.I.R. Khan, P.S. Reddy, A. Ferrante and N.A. Khan), 437–448. Woodhead Publishing.

Jahan, M.S., Wang, Y., Shu, S. et al. (2019). Exogenous salicylic acid increases the heat tolerance in tomato (*Solanum lycopersicum* L) by enhancing photosynthesis efficiency and improving antioxidant defense system through scavenging of reactive oxygen species. *Scientia Horticulturae* 247: 421–429.

Janeczko, A., Oklešťková, J., Pociecha, E. et al. (2011). Physiological effects and transport of 24-epibrassinolide in heat-stressed barley. *Acta Physiologiae Plantarum* 33 (4): 1249–1259.

Kaushal, N., Bhandari, K., Siddique, K.H., and Nayyar, H. (2016). Food crops face rising temperatures: an overview of responses, adaptive mechanisms, and approaches to improve heat tolerance. *Cogent Food & Agriculture* 2 (1): 1134380.

Kazan, K. (2013). Auxin and the integration of environmental signals into plant root development. *Annals of Botany* 112: 1655–1665.

Khalil, U., Ali, S., Rizwan, M. et al. (2018). Role of mineral nutrients in plant growth under extreme temperatures. In: *Plant Nutrients and Abiotic Stress Tolerance* (eds. M. Hasanuzzaman, M. Fujita, H. Oku, et al.), 499–524. Singapore: Springer.

Koini, M.A., Alvey, L., Allen, T. et al. (2009). High temperature-mediated adaptations in plant architecture require the bHLH transcription factor PIF4. *Current Biology* 19 (5): 408–413.

Kosar, F., Akram, N.A., Sadiq, M. et al. (2018). Trehalose: a key organic osmolyte effectively involved in plant abiotic stress tolerance. *Journal of Plant Growth Regulation*: https://doi.org/10.1007/s00344-018-9876-x: 1–13.

Kumar, S.V., Lucyshyn, D., Jaeger, K.E. et al. (2012). Transcription factor PIF4 controls the thermosensory activation of flowering. *Nature* 484 (7393): 242.

Kumar, R.R., Sharma, S.K., Goswami, S. et al. (2015). Salicylic acid alleviates the heat stress-induced oxidative damage of starch biosynthesis pathway by modulating the expression of heat-stable genes and proteins in wheat (*Triticum aestivum*). *Acta Physiologiae Plantarum* 37 (8): 143.

Larkindale, J. and Huang, B. (2005). Effects of abscisic acid, salicylic acid, ethylene and hydrogen peroxide in thermotolerance and recovery for creeping bentgrass. *Plant Growth Regulation* 47 (1): 17–28.

Larkindale, J. and Knight, M.R. (2002). Protection against heat stress-induced oxidative damage in *Arabidopsis* involves calcium, abscisic acid, ethylene, and salicylic acid. *Plant Physiology* 128 (2): 682–695.

Li, Z.G. and Jin, J.Z. (2016). Hydrogen sulfide partly mediates abscisic acid-induced heat tolerance in tobacco (*Nicotiana tabacum* L.) suspension cultured cells. *Plant Cell, Tissue and Organ Culture* 125 (2): 207–214.

Li, Z.G., Gong, M., Xie, H. et al. (2012). Hydrogen sulfide donor sodium hydrosulfide-induced heat tolerance in tobacco (*Nicotiana tabacum* L.) suspension cultured cells and involvement of Ca^{2+} and calmodulin. *Plant Science* 185: 185–189.

Li, Z.G., Luo, L.J., and Zhu, L.P. (2014). Involvement of trehalose in hydrogen sulfide donor sodium hydrosulfide-induced the acquisition of heat tolerance in maize (*Zea mays* L.) seedlings. *Botanical Studies* 55 (1): 1–9.

Li, Z.G., Xie, L.R., and Li, X.J. (2015). Hydrogen sulfide acts as a downstream signal molecule in salicylic acid-induced heat tolerance in maize (*Zea mays* L.) seedlings. *Journal of Plant Physiology* 177: 121–127.

Lin, K.H., Huang, S.B., Wu, C.W., and Chang, Y.S. (2019). Effects of salicylic acid and calcium chloride on heat tolerance of *Poinsettia*. *Hort Science* 54 (3): 499–504.

Liu, Z.M., Yue, M.M., Yang, D.Y. et al. (2017). Over-expression of SlJA2 decreased heat tolerance of transgenic tobacco plants via salicylic acid pathway. *Plant Cell Reports* 36 (4): 529–542.

Lobell, D.B., Schlenker, W., and Costa-Roberts, J. (2011). Climate trends and global crop production since 1980. *Science* 333 (6042): 616–620.

Luo, Y., Li, F., Wang, G.P. et al. (2010). Exogenously-supplied trehalose protects thylakoid membranes of winter wheat from heat-induced damage. *Biologia Plantarum* 54: 495–501.

Mazorra, L.M., Nunez, M., Hechavarria, M. et al. (2002). Influence of brassinosteroids on antioxidant enzymes activity in tomato under different temperatures. *Biologia Plantarum* 45 (4): 593–596.

Min, L., Li, Y., Hu, Q. et al. (2014). Sugar and auxin signaling pathways respond to high-temperature stress during anther development as revealed by transcript profiling analysis in cotton. *Plant Physiology* 164: 1293–1308.

Minocha, R., Majumdar, R., and Minocha, S.C. (2014). Polyamines and abiotic stress in plants: a complex relationship. *Frontiers in Plant Science* 5: 175.

Mishra, S., Srivastava, V., Mehrotra, S., and Quadri, S.N. (2017). The regulation of plant development: cross-talk of reactive oxygen species and plant hormones. In: *Reactive Oxygen Species in Plants: Boon or Bane:Revisiting the Role of ROS* (eds. V.P. Singh, S. Singh, D.K. Tripathi, et al.). Chichester: Wiley https://doi.org/10.1002/9781119324928.ch13.

Nahar, K., Hasanuzzaman, M., and Fujita, M. (2016). Roles of osmolytes in plant adaptation to drought and salinity. In: *Osmolytes and Plants Acclimation to Changing Environment: Emerging Omics Technologies* (eds. N. Iqbal, R. Nazar and N.A. Khan), 37–68. New Delhi: Springer.

Nazar, R., Iqbal, N., and Umar, S. (2017). Heat stress tolerance in plants: action of salicylic acid. In: *Salicylic Acid: A Multifaceted Hormone* (eds. R. Nazar, N. Iqbal and N.A. Khan), 145–161. Singapore: Springer.

Ogweno, J.O., Song, X.S., Shi, K. et al. (2008). Brassinosteroids alleviate heat-induced inhibition of photosynthesis by increasing carboxylation efficiency and enhancing antioxidant systems in *Lycopersicon esculentum*. *Journal of Plant Growth Regulation* 27 (1): 49–57.

Ozga, J.A., Kaur, H., Savada, R.P., and Reinecke, D.M. (2016). Hormonal regulation of reproductive growth under normal and heat-stress conditions in legume and other model crop species. *Journal of Experimental Botany* 68 (8): 1885–1894.

Parankusam, S., Adimulam, S.S., Bhatnagar-Mathur, P., and Sharma, K.K. (2017). Nitric oxide (NO) in plant heat stress tolerance: current knowledge and perspectives. *Frontiers in Plant Science* 8: 1582.

Rasheed, R., Wahid, A., Farooq, M. et al. (2011). Role of proline and glycinebetaine pretreatments in improving heat tolerance of sprouting sugarcane (*Saccharum* sp.) buds. *Plant Growth Regulation* 65: 35–45.

Rezaul, I.M., Baohua, F., Tingting, C. et al. (2019). Abscisic acid prevents pollen abortion under high-temperature stress by mediating sugar metabolism in rice spikelets. *Physiologia Plantarum* 165 (3): 644–663.

Sagor, G.H.M., Berberich, T., Takahashi, Y. et al. (2013). The polyamine spermine protects *Arabidopsis* from heat stress-induced damage by increasing expression of heat shock-related genes. *Transgenic Research* 22 (3): 595–605.

Sakata, T., Oshino, T., Miura, S. et al. (2010). Auxins reverse plant male sterility caused by high temperatures. *Proceedings of the National Academy of Sciences* 107: 8569–8574.

Salomé, P.A. (2017). Some like it HOT: protein translation and heat stress in plants. *The Plant Cell* 29: 2075.

Senaratna, T., Touchell, D., Bunn, E., and Dixon, K. (2000). Acetyl salicylic acid (aspirin) and salicylic acid induce multiple stress tolerance in bean and tomato plants. *Plant Growth Regulation* 30 (2): 157–161.

Sequera-Mutiozabal, M., Antoniou, C., Tiburcio, A.F. et al. (2017). Polyamines: emerging hubs promoting drought and salt stress tolerance in plants. *Current Molecular Biology Reports* 3 (1): 28–36.

Sergiev, I., Todorova, D., Shopova, E. et al. (2018). Effects of auxin analogues and heat stress on garden pea. *Zemdirbyste-Agriculture* 105 (3): 243–248.

Sharma, M. and Laxmi, A. (2016). Jasmonates: emerging players in controlling temperature stress tolerance. *Frontiers in Plant Science* 6: 1129.

Sharma, L., Dalal, M., Verma, R.K. et al. (2018). Auxin protects spikelet fertility and grain yield under drought and heat stresses in rice. *Environmental and Experimental Botany* 150: 9–24.

Shi, H. and Chan, Z. (2014). Improvement of plant abiotic stress tolerance through modulation of the polyamine pathway. *Journal of Integrative Plant Biology* 56 (2): 114–121.

Shi, Q., Bao, Z., Zhu, Z. et al. (2006). Effects of different treatments of salicylic acid on heat tolerance, chlorophyll fluorescence, and antioxidant enzyme activity in seedlings of *Cucumis sativa* L. *Plant Growth Regulation* 48 (2): 127–135.

Siddiqui, H., Hayat, S., and Bajguz, A. (2018). Regulation of photosynthesis by brassinosteroids in plants. *Acta Physiologiae Plantarum* 40 (3): 59.

Skoog, F. and Miller, C.O. (1957). Chemical regulation of growth and organ formation in plant tissues cultured *in vitro*. *Symposia of the Society for Experimental Biology* 54: 118–130.

Song, L., Ding, W., Zhao, M. et al. (2006). Nitric oxide protects against oxidative stress under heat stress in the calluses from two ecotypes of reed. *Plant Science* 171 (4): 449–458.

Song, L., Jiang, Y., Zhao, H., and Hou, M. (2012). Acquired thermotolerance in plants. *Plant Cell, Tissue and Organ Culture* 111 (3): 265–276.

Suzuki, N., Bassil, E., Hamilton, J.S. et al. (2016). ABA is required for plant acclimation to a combination of salt and heat stress. *PLoS One* 11 (1): e0147625.

Thussagunpanit, J., Jutamanee, K., Kaveeta, L. et al. (2015a). Comparative effects of brassinosteroid and brassinosteroid mimic on improving photosynthesis, lipid peroxidation, and rice seed set under heat stress. *Journal of Plant Growth Regulation* 34 (2): 320–331.

Thussagunpanit, J., Jutamanee, K., Sonjaroon, W. et al. (2015b). Effects of brassinosteroid and brassinosteroid mimic on photosynthetic efficiency and rice yield under heat stress. *Photosynthetica* 53 (2): 312–320.

Todorova, D., Genkov, T., Vaseva-Gemisheva, I. et al. (2005). Effect of temperature stress on the endogenous cytokinin content in *Arabidopsis thaliana* (L.) Heynh plants. *Acta Physiologiae Plantarum* 27 (1): 13–18.

Vierling, E. (1991). The roles of heat shock proteins in plants. *Annual Review of Plant Biology* 42 (1): 579–620.

Wang, L.J., Fan, L., Loescher, W. et al. (2010). Salicylic acid alleviates decreases in photosynthesis under heat stress and accelerates recovery in grapevine leaves. *BMC Plant Biology* 10 (1): 34.

Wu, J.Z., Lin, Y., Zhang, X.L. et al. (2008). IAA stimulates pollen tube growth and mediates the modification of its wall composition and structure in *Torenia fournieri*. *Journal of Experimental Botany* 59 (9): 2529–2543.

Wu, C., Cui, K., Wang, W. et al. (2016). Heat-induced phytohormone changes are associated with disrupted early reproductive development and reduced yield in rice. *Scientific Reports* 6: 34978.

Wu, C., Cui, K., Wang, W. et al. (2017). Heat-induced cytokinin transportation and degradation are associated with reduced panicle cytokinin expression and fewer spikelets per panicle in rice. *Frontiers in Plant Science* 8: 371.

Yang, D., Li, Y., Shi, Y. et al. (2016). Exogenous cytokinins increase grain yield of winter wheat cultivars by improving stay-green characteristics under heat stress. *PLoS One* 11 (5): e0155437.

Yoshida, S., Satake, T., and Mackill, D.S. (1981). High-temperature stress in rice (study conducted at IRRI, Philippines). *IRRI Research Paper Series (Philippines)* 67: 1–15.

Zhang, C.X., Feng, B.H., Chen, T.T. et al. (2017). Sugars, antioxidant enzymes and IAA mediate salicylic acid to prevent rice spikelet degeneration caused by heat stress. *Plant Growth Regulation* 83 (2): 313–323.

Zhang, C., Li, G., Chen, T. et al. (2018). Heat stress induces spikelet sterility in rice at anthesis through inhibition of pollen tube elongation interfering with auxin homeostasis in pollinated pistils. *Rice* 11 (1): 14.

11

Exploring In-built Defense Mechanisms in Plants under Heat Stress

Giridara Kumar Surabhi and Jatindra Kumar Seth

Plant Molecular Biology and 'OMICS' Laboratory, Regional Plant Resource Centre, Bhubaneswar, Odisha, India

11.1 Introduction

Global warming is predicted to be accompanied by more frequent and powerful extreme temperature events. Therefore, a better understanding of the response of crop plants to heat stress is essential for improving their heat tolerance. At very high temperatures, a catastrophic collapse of cellular organization may occur within minutes, which results in severe cellular damage and even cell death (Tubiello et al. 2007; Ding et al. 2018). Every crop species with accord to their physiology has their own range of inbuilt temperature *maxima* and *minima* at different developmental stages, beyond which all these processes are inhibited (Kaushal et al. 2016). The impact of global climatic changes resulting from the constant elevation in atmospheric carbon dioxide (CO_2) levels is exacerbated by the constant decline in agricultural land availability and quality (Peters et al. 2011). Relative to 1850–1900, the global surface temperature change is projected to exceed 2 °C by the end of the twenty-first century as revealed by the Intergovernmental Panel on Climate Change 2014 (Zandalinas et al. 2018). Actually, plants are not passively damaged by heat stress, but respond to temperature changes by reprogramming their composition of certain transcripts, proteins, and metabolites. Such changes are aimed at establishing a new steady-state balance of biological

Heat Stress Tolerance in Plants: Physiological, Molecular and Genetic Perspectives, First Edition.
Edited by Shabir Hussain Wani and Vinay Kumar.

processes that enable the organism to function, survive, and even reproduce at a higher temperature.

High temperature (HT) stress inhibits the redox and metabolic balance in photosystem II (PSII), photosystem I (PSI), and Rubisco (ribulose1,5-bisphosphate carboxylase/oxygenase) of the photosynthetic apparatus and increase water loss through transpiration, reduce pollen viability and grain filling, resulting in marked yield loss (Higashitani 2013; Mathur et al. 2014; Cheng et al. 2017; Rieu et al. 2017). Thus, agricultural production around the globe is threatened by HT stress (Feller 2016; Cheng et al. 2017). HT stress can be expected to become increasingly problematic as global warming leads to more adverse climatic changes (Tubiello et al. 2007). In addition, it is expected that crops in future climates will be exposed not only to higher mean air temperatures but also to more frequent short episodes of HT stress (Djanaguiraman et al. 2018).

Plants have evolved a variety of response mechanisms to elevated temperatures and the mechanisms of plant responses to HT stress vary with plant type and the degree and duration of HT (Kotak et al. 2007; Wang et al. 2018). Plants synthesizes compatible solutes that modify proteins and cellular structures, sustain cell turgor by osmotic adjustment, increase cyclic electron flow in photosynthetic systems, and regulate the antioxidant system to re-establish the cellular redox balance and homeostasis (Xiong and Zhu 2001; Antunes et al. 2013; Driedonks et al. 2016; Cheng et al. 2017). Many HT-responsive genes have been identified and altered gene expression plays an important role in plant HT tolerance (Larkindale and Vierling 2008; Mittler et al. 2012; Ding et al. 2018). The elucidation of molecular mechanisms underlying the persistence of plants to HT stress and how they respond and adapt to HT may facilitate in the development of novel breeding strategies to generate HT-tolerant plants (Ohama et al. 2016; Wang et al. 2018). This chapter focuses on different inbuilt mechanisms in plants to cope with HT stress in crop plants.

11.2 Effect of Heat Stress on Crop Plants

HT stress targets plant on various stages of their development and limits their growth. It can cause a number of morphological, physiological, and biochemical changes in plants, which leads to drastic reductions in crop yield. Heat stress on plants induces the accumulation of reactive oxygen species (ROS) (Zandalinas et al. 2018), which causes cellular damage and inhibition of photosynthesis (Ahanger et al. 2017). Heat stress targets the reproductive stage as it is more sensitive to heat and causes impaired fertilization and abortion of flowers (Kaushal et al. 2016). This impaired reproductive growth is followed by the inhibition of pollen grain swelling and anther dehiscence, which affects seed production (Das et al. 2014). The effect of HT on plants can be studied under four subheadings with respect to their targets, as it affects different physiological aspects, cell structures, vegetative growth, and reproductive growth.

11.2.1 Heat Stress Effects on Physiology and Cell Structures

Photosynthesis is one of the major thermosensitive processes (Kim and Portis 2005; Kaushal et al. 2016). Light reaction is reduced as heat targets PSII and in dark reaction it affects the activity of Rubisco (Ahmad et al. 2010; Kaushal et al. 2016; Martinez et al. 2018). Heat stress regulates mitochondrial activity of the crop by which it affects the plant respiration (Stone 2001; Kaushal et al. 2016). With the increase in temperature the rate of respiration increases, but beyond threshold levels respiration decreases, owing to damage to respiratory machinery (Prasad et al. 2008). Decreased respiration has been reported in chickpea under HT (Kumar et al. 2013). This was most likely due to the effect of impaired structure and function of mitochondria, proteins, and electron transport chain (Kaushal et al. 2016). The HT stress positively regulates the rate of both photorespiration and dark respiration in cotton (Salvucci and Crafts-Brandner 2004), and its effects in physiology of crops has been reported on different crops such as wheat (Prasad et al. 2011; Mishra et al. 2017), sorghum (Djanaguiraman et al. 2010, 2017), tomato (Martinez et al. 2018), grape (Jiang et al. 2017), and lablab (Rai et al. 2018). HT injury causes loss in pre-harvest and post-harvest aspects of crop (Almeselmani et al. 2006).

Cell structures like plasma membranes are heat-sensitive as they are the primary sites for injury among all components of a plant cell (Kaushal et al. 2016). HT causes the plants to undergo a phase transition from solid–gel structure to flexible liquid–crystalline structure of the plasma membrane. The membrane proteins are denatured and unsaturated fatty acids concentration increases for which the fluidity of the membrane is increased (Savchenko et al. 2002). Membrane injury has been observed in sensitive chickpea genotypes at 40/30 °C which was further aggravated at 45/35 °C (day and night temperatures) (Kumar et al. 2013).

11.2.2 Heat Stress Effects on Vegetative Stages

Heat stress can affect the rate of growth and development of plants in accordance to the intensity, duration, and stage of exposure (Gan et al. 2004; Wahid and Close 2007; Kaushal et al. 2016). Seeds in their developing stages are more sensitive to temperature, and long-term exposure of heat may cause delayed germination of seeds or loss of vigor, which ultimately reduces seedling establishment (Djanaguiraman et al. 2018). Reduction in seed germination under HT stress was studied in many legumes including lentil (Sehgal et al. 2018), mung bean (Devasirvatham et al. 2012; Kumar et al. 2011), chickpea (Piramila et al. 2012), and in other crop plants, such as millet (Djanaguiraman et al. 2018) and grape (Jiang et al. 2017).

Plants can show numerous morphological symptoms in heat stress condition such as scorching and sun-burning of leaves, twigs, branches and stems, leaf senescence and abscission, shoot and root growth inhibition, fruit discoloration, and damage which ultimately reduces yield (Vollenweider and Günthardt-Goerg 2005). For example, decreased plant height under heat stress has been reported in wheat (Laghri et al. 2012) and alfalfa (Mingpeng et al. 2010). Inhibition of leaf growth with long exposures to heat stress was observed in wheat (Savicka and Škute 2010). Reductions in shoot dry mass, growth, and net assimilation rate due to HT has been elucidated in wheat (Prasad et al. 2011), millet (Djanaguiraman et al. 2018), *Pyropia* (Shi et al. 2017), and sorghum (Djanaguiraman et al. 2017). Heat stress causing chlorosis at vegetative growth stage has been reported in maize (Karim et al. 1997), wheat (Almeselmani et al. 2009), mung bean (Kumar et al. 2011), and chickpea (Kumar et al. 2013).

11.2.3 Heat Stress Effects on Reproductive Stage

The reproductive stage of many crop species has been affected by heat stress and suffers loss of buds, flowers, fruit, and seeds resulting in severe reductions in crop productivity (Kaushal et al. 2016; Prasad et al. 2017; Djanaguiraman et al. 2018). Crop flowering during heat stress causes reduction in flower number and size, and deformed floral organs leading to loss of flowers and young pods and in turn reduces yield (Kaushal et al. 2016; Djanaguiraman et al. 2018; Narayanan et al. 2018). Damage to male or female gametophytes is determined at micro sporogenesis or mega sporogenesis stages, respectively (Devasirvatham et al. 2012; Nayyar et al. 2005a,b; Prasad et al. 2017). Recently, HT stress effects on female gametophyte of chickpea (Devasirvatham et al. 2013; Kaushal et al. 2016) and tomato (Giorno et al. 2013) have been studied. The number and morphology of pollen grains, pollen viability, metabolism, and composition are affected by heat stress, in wheat (Narayanan et al. 2018; Djanaguiraman et al. 2018), lentil (Sehgal et al. 2018), sorghum (Djanaguiraman et al. 2018), and chickpea (Devasirvatham et al. 2013). It was noticed that female gametophytic tissue is less sensitive as compared to the male gametophytic tissue toward heat stress (Devasirvatham et al. 2013; Kaushal et al. 2013). The pistil is sporophytic tissue which provides nutrition to developing male and female gametophytes; heat stress damages its functions, which causes impaired tube growth (Snider et al. 2011). The receptivity of stigma reduces significantly at 40/30 and 45/35 °C in chickpea (Kumar et al. 2013). HT stress obstructs fertilization by inhibiting the development of male and female gametophyte (Kaushal et al. 2016).

11.2.4 Heat Stress Effects on Yield

Heat stress causes reduced yield in legumes, cereals, and other crops, owing to improper vegetative and reproductive development (Kaushal et al. 2016; Prasad et al. 2017). Reductions in yield has been reported in many food crops like rice (Islam 2011), wheat (Zhang et al. 2013; Green et al. 2018), soybean (Board and Kahlon 2011), lentil (Barghi et al. 2012), cowpea (Hall 1992), and chickpea (Kumar et al. 2013), owing to reductions in seed filling.

Heat tolerance of the plants at the reproductive stage can be determined by deriving the duration of grain filling (Green et al. 2018). Another stud suggests that an increase in dark reaction under heat stress condition leads to reduced yield in wheat due to the production of immature grains (Johkan et al. 2011). Heat stress (15/10 and 21/16 °C day/night) during the stage of grain filling resulted in smaller grains, which further reduces yield in wheat (Maçãs et al. 2000). Seed yield and harvest index of chickpea was reduced with exposure to HTs during flower and pod development (Gan et al. 2004; Kumar et al. 2013; Wang et al. 2006). Reduced grain yield in maize under heat stress is reported by Khodarahmpour (2011). Fruit set is one of the vulnerable stages and is severely affected by heat stress (Prasad et al. 2017). Many studies have reported reduction in fruit set, such as in cowpea (Ahmed et al. 1993), cotton (Reddy et al. 1999), groundnut (Prasad et al. 2003), and field pea (Jiang et al. 2018).

11.3 Threshold Temperature

Plants have different tolerance capacity toward temperature; hence threshold temperature is defined as the temperature beyond which plants do not resist it and their growth and development are affected, sometimes leading to death of the crop (Kaushal et al. 2016; Prasad et al. 2018). For each crop there is a minimal temperature at which crops can grow easily, i.e. T_{min}, but below this plant growth stops, and maximum temperature, i.e. T_{max}, above which growth stops, respectively (Prasad et al. 2018). The temperature at which a plant shows its maximum growth and development is called optimal temperature, i.e. T_{opt} (Prasad et al. 2018). Different crops have their own threshold temperature, which is the optimal temperature for their vegetative and reproductive growth (Hatfield et al. 2011; Zinn et al. 2010). When the threshold temperature for a crop species exceeds, its seed germination, seedling establishment, and vegetative growth, flowering, fruit set, and fruit ripening are adversely affected (Kaushal et al. 2016).

11.4 In-built Defense System in Plants to Overcome High Temperature Stress

Heat tolerance is generally induced in crop plants by exposing them to moderate HTs, which enables the plant to modify its mechanism or adapt to certain environments and this acclimatization is termed as thermotolerance (Kaushal et al. 2016). Various plant metabolic pathways such as ROS regulatory systems; hormone signaling, including abscisic acid (ABA), salicylic acid (SA), jasmonic acid (JA), ethylene signaling; as well as heat shock proteins (HSPs) dependent pathways alter in plants in order to protect against HT stress (Suzuki and Katano 2018). The molecules involved in stress protection are regarded as thermoprotectants (Kaushal et al. 2016). In addition, in the process of adaptation of thermotolerance, plants accumulate various compounds or compatible solutes such as proline (Pro), glycinebetaine (GB), trehalose, brassinosteroids (BRs), and polyamines (Pas), which play an important role in thermotolerance (Kaushal et al. 2016). For instance, plants can produce solutes that modify proteins and cellular structures, sustain cell turgor by osmotic adjustment, increase cyclic electron flow in photosynthetic systems, and regulate efficient antioxidant system to re-establish the cellular redox balance and homeostasis (Xiong and Zhu 2001; Antunes et al. 2013; Driedonks et al. 2016). To date, a very few studies have focused on the involvement of such molecules in heat-stress tolerance and some of these are discussed in this chapter.

11.4.1 Accumulation of Thermoprotectants

11.4.1.1 Heat Shock Proteins (HSPs)

An essential component of acquired thermotolerance in plants and other organisms is the induction and synthesis of molecular chaperones or HSPs, and many of the molecular chaperones were originally identified as HSPs (Kaushal et al. 2016; Ohama et al. 2016; Shi et al. 2017; Suzuki and Katano 2018; Wang et al. 2018; Zandalinas et al. 2018). HSPs play a vital role in protecting plant cells from the deleterious effects of heat stress (Rousch et al. 2004). Under heat stress, the synthesis of transcription and translation of HSPs increases manifold (Kaushal et al. 2016). HSPs protect the stability and functional confirmation of proteins through folding and re-folding of non-native proteins under stress conditions (Tripp et al. 2009; Wang et al. 2018). HSPs are found in all groups of living organisms and are classified into five different types according to their molecular weight: HSP100, HSP90, HSP70, HSP60, the chaperonins (GroEL), and small heat shock proteins (sHSPs) (Ohama et al. 2016; Shi et al. 2017; Zandalinas et al. 2018). Genome analysis revealed the existence of approximately 13 sHSPs, seven HSP60, 18 HSP70, seven HSP90, and eight HSP100 members in *Arabidopsis thaliana*. HSPs have enhanced thermotolerance in different crop plants such as soybean (Ortiz and Cardemil 2001), maize (Queitsch et al. 2000), tomato (Frank et al. 2009), grape (Jiang et al. 2017), and *Pyropia* (Wang et al. 2018). Shi et al. (2017) report that up-regulation of HSPs prevents denaturation of other proteins caused by HT, and thus improves the HT tolerance of *Pyropia*. HSP70 has been observed as the most abundant under HT stress in crops, indicating its fundamental role in response to HT stress (Wang et al. 2018). HSPs are reviewed extensively in the literature (Hasanuzzaman et al. 2013; Larkindale and Knight 2002; Wang et al. 2004; Kaushal et al. 2016). It has been found that HSP70, HSP90, and the heat shock transcription factors (HSF) HSFA2 and HSFA3 were important in tomato for enhancing microspore resistance to heat stress (Frank et al. 2009; Jiang et al. 2017). Individually, each member of HSPs has a unique function, but the coordinated function of different HSPs networks appears as a key for heat-stress tolerance. However, the role of HSPs/chaperones as regulatory and sensing molecules in signal transduction and transcription pathways during stress remains unclear. Hence the role of HSPs accumulation and its metabolic networks vis-à-vis tolerance to HT needs to be critically examined.

11.4.1.2 Proline

Several studies conducted earlier revealed the capacity to accumulate Proline (Pro) has been correlated with stress tolerance

(Surabhi et al. 2000, 2003). Pro, a nonessential amino acid, which has an amino group (–NH) instead of the usual amino group (–NH_2), is one of the most studied thermoprotectant (Kaushal et al. 2016). Pro is an amino acid which plays a vital role in stress condition, acts as an excellent osmolyte, and/or a signaling molecule (Farag et al. 2018). Biosynthesis of Pro in response to heat stress and the involvement of nitric oxide (NO) and Pro in thermotolerance mechanism in faba bean (*Vicia faba* L.) has been reported (Alamri et al. 2018). Under thermal stress condition, accumulation of Pro in the cotyledons of watermelon was observed, suggesting its crucial role in heat tolerance (Silva et al. 2018). Having protective or scavenging features, it has shown that Pro metabolism can stabilize cellular homeostasis during stress conditions which are not yet properly understood (Murmu et al. 2017). Many studies have revealed that there is a positive relationship between the accumulation of Pro and plant stress tolerance; however, some argue that increased Pro concentration under stress is a product and not an adaptive response to stress (Surabhi et al. 2000, 2003; Ashraf and Foolad 2007; Kaushal et al. 2016). Pro acts as a molecular chaperone and regulates protein integrity and enhances the activities of different enzymes. Its functions include the prevention of protein aggregation and stabilization of M4 lactate dehydrogenase during extreme temperatures (Sharma and Dubey 2005; Murmu et al. 2017). Many studies have attributed an antioxidant feature to Pro, suggesting ROS-scavenging activity of Pro, and it acts as a singlet oxygen quencher (Smirnoff and Cumbes 1989; Murmu et al. 2017). Under HT stress condition (45/40 °C) minimal injury to membranes, improved chlorophyll and water contents was observed in different crop species including chickpea (Kaushal et al. 2016). Pro accumulation under heat stress was reported in wheat (Ahmed and Hasan 2011), maize (Kumar et al. 2012), tobacco (Cvikrová et al. 2012), watermelon (Silva et al. 2018), faba bean (*V. faba*) (Siddiqui et al. 2015; Alamri et al. 2018), moth bean (*Vigna aconitifolia*) (Harsh et al. 2016), common bean (Chavez-Arias et al. 2018), and soft corals (Farag et al. 2018). In contrast, no Pro accumulation has been observed in *Arabidopsis* (Rizshky et al. 2004), while a slight decrease was observed in germinating wheat seeds under heat stress (Song et al. 2005; Kaushal et al. 2016).

11.4.1.3 Glycinebetaine (GB)

When the crop plants are exposed to abiotic stress factors, they are stimulated to accumulate certain organic compounds of low molecular mass in the cytoplasm. These organic compounds are collectively known as "compatible solutes" (Surabhi et al. 2000). Among different compatible solutes, one of the most extensively studied osmoprotectant is GB. Many taxonomically distant plant species normally contain low levels of GB (these plants are known as natural accumulators of GB), but they can also accumulate larger amounts of GB when subjected to abiotic stress. In many other species, GB has not been detected under normal or stressful conditions (natural nonaccumulators) (Chen and Murata 2011). Levels of GB accumulation vary considerably among plant species and organs (Chen and Murata 2011). GB is a low molecular mass compatible solute which is a soluble quaternary ammonium compound without cellular toxicity (Wyn Jones and McDonnell 1988; Cleland et al. 2004). It is dipolar but electrically neutral at physiological pH (Rhodes and Hanson 1993).

GB effectively stabilizes the quaternary structures of enzymes and complex proteins, protects various components of the photosynthetic machinery (Shahbaz et al. 2011), such as Rubisco, and the oxygen-evolving PSII (Murata et al. 1992; Rhodes and Hanson 1993; Lee et al. 1997; McNeil et al. 1999; Nayyar et al. 2005a,b), maintains highly ordered state of membranes at nonphysiological temperatures (Zhao et al. 1992; Rajashekar 2000; Sakamoto and Murata 2002; Nayyar et al. 2005a,b; Fariduddin et al. 2013) and mitigate oxidative damage (Chen and Murata 2011) and high salt concentrations (Papageogiou and Murata 1995).

Under HT stress, crops have been seen to accumulate GB to stimulate synthesis of the D1 protein, which supports the repair of photodamaged PSII (Chen et al. 2011; Allakhverdiev et al. 2007). Further, accumulation of GB prevents the seizure of Rubisco activase to the thylakoid membrane, thereby maintaining the activity of Rubisco at HT (Yang et al. 2005). Moreover, it has been found that the accumulation of GB improves the inhibition of net photosynthetic rate under heat stress (Wang et al. 2010).

However, a precise role of compatible solutes, including GB, in abiotic stress tolerance is largely unknown (Giri 2011). With the increasing efforts of whole genome sequencing for several crop plants coupled with recent advances in high-throughput omics approaches and genome editing can shed light on understanding GB biosynthetic pathway more thoroughly toward heat-stress tolerance.

11.4.1.4 Abscisic Acid (ABA)

Hormones (such as ABA), secondary messengers, phosphatases, and protein kinases are crucial components within the stress-induced signaling network that regulates a multitude of biochemical and physiological processes (Hirayama and Shinozaki 2010). ABA is an integral regulator of abiotic stress signaling (Raghavendra et al. 2010). ABA quickly accumulates in response to different environmental stress conditions, and it has been observed that ABA-deficient plants have an altered stress response. ABA promotes stomatal closure, inhibits stomatal opening to reduce water loss by transpiration, and induces the expression of numerous stress-related genes, and some studies indicate a role in the regulation of stress-induced metabolic adjustments (Krasensky and Jonak 2012).

ABA is a naturally occurring plant hormone that helps to regulate plant growth and development. Despite reports of increased ABA levels in plants exposed to HT, little is known about ABA accumulation in plants during heat acclimation (Penfield 2008). Many studies have reported the accumulation of ABA under heat stress, such as in maize (Gong et al. 1998), bromegrass (*Bromus inermis* L; Robertson et al. 1994), tobacco (Teplova et al. 2000), pepper (Tong-Xiang et al. 2009), mustard (Chhabra et al. 2009), and *Arabidopsis* (Suzuki et al. 2014). ABA is a signaling molecule, but it is reported to induce thermotolerance by raising the level of other signaling molecules, like NO, as observed in common/giant reed (Song et al. 2008; Kaushal et al. 2016). It has been reported that when salt stress was combined with HT, an increase in ABA levels was observed. This was thought to counteract the aperture of stomata caused by HT in citrus plants (Balfagón et al. 2018). ABA also induces the accumulation of different proteins involved in the acclimation and regulates stomatal closure under heat stress (Finkelstein 2013; Gómez-Cadenas et al. 2015). In *Arabidopsis*, it was observed that ABA plays an important role in accumulating essential proteins during interactive effect of water deficit and heat stress (Zandalinas et al. 2016). The mutants deficient in ABA metabolism and signaling were more susceptible than wild-type plants under combination of salt and heat stress (Suzuki et al. 2014). The catabolic product of ABA, phaseic acid (PA) acts as a signaling molecule that fine-tunes environmental adaptation and development (Balfagón et al. 2018) in crop plants. Interestingly, accumulation of many osmotic stress-induced proteinogenic amino acids, including Pro, were induced and depended on ABA (Urano et al. 2009).

11.4.1.5 Salicylic Acid (SA)

SA is an important endogenous signaling hormone induced under heat stress conditions in plants. Under heat stress plants accumulate a range of SA (1 μM–1 mM) and its levels increases with intensity of heat stress (Kaushal et al. 2016). Its role in heat tolerance was studied in mustard seedlings (Dat et al. 1998). An increase in endogenous SA levels in response to heat was reported in pea plants (Pan et al. 2006; Kaushal et al. 2016). With the accumulation of SA, plants increase their stress tolerance capacity toward the combination of both drought and HT (Zandalinas et al. 2016). SA is a phenolic phytohormone which regulates thermo tolerance mechanisms such as maintenance of membrane integrity and protection of PSII complex in crop plants (Wang et al. 2010; Wang et al. 2014; Zandalinas et al. 2017). Recent study has revealed that the highest hormone content as well as up-regulation of *ICS* (SA-biosynthetic gene) and *PR2* (a SA-inducible gene) in plants subjected to combined moderate drought and heat stress (Dempsey et al. 2011; Coqueiro et al. 2015; Zandalinas et al. 2017). In addition, the remarkable SA accumulation and increased signaling during heat stress was noticed in citrus, this could be due to sensitivity of citrus genotype to HT, and aimed to reduce the PSII damage and protect against the different oxidative damages (Zandalinas et al. 2017). However, more focus is required on SA and its role in HT protection in plants.

Antioxidant Defense System HT stress causes the production of ROS such as hydrogen peroxide (H_2O_2), superoxide radical (O^{2-}), singlet oxygen ($^1O_2$), and hydroxyl radical (OH^-), which causes oxidative damage to plant tissue (Ohama et al. 2016; Djanaguiraman et al. 2018). To cope with this oxidative damage plants possess an antioxidant enzyme defense system, which includes enzymes such as superoxide dismutase (SOD), catalase (CAT), and ascorbate peroxidase (APX) (Sudhakar et al. 2001; Ohama et al. 2016; Djanaguiraman et al. 2018; Wang et al. 2018). The antioxidant system scavenges ROS, as SOD is a major scavenger of superoxide ($O_2^{\cdot -}$) and its enzymatic action results in the formation of H_2O_2 and O_2. Catalase and a variety of peroxidases catalyze the breakdown of H_2O_2. Catalase, which is apparently absent in the chloroplast, dismutases H_2O_2 into water and molecular oxygen, whereas peroxidase decomposes H_2O_2 by oxidation of co-substrates such as phenolic compounds and/or antioxidants (Sudhakar et al. 2001; Wang et al. 2018; Djanaguiraman et al. 2018).

Under HT the catalase unigenes are up-regulated in *Pyropia* (Wang et al. 2018). Another antioxidant, thioredoxin (TRX) also protects cells from oxidative damage caused by HT stress (Shi et al. 2018). The efficient antioxidants toward stress protection was reported in many studies such as wheat (Djanaguiraman et al. 2018), tomato (Ding et al. 2018), lablab (Rai et al. 2018), millet (Djanaguiraman et al. 2018), and radish (Wang et al. 2018). The antioxidant enzymes activity in chloroplasts was down-regulated by HT stress, exacerbating accumulation of ROS in wheat (Djanaguiraman et al. 2018). Three unigenes (Unigene8530, Rsa#S4302246, and EX767247), one unigene (CL7724.Contig1), and three unigenes (Unigene13837, CL11601.Contig2, and CL11601.Contig1) encoding Cu/Zn-SOD, Fe-SOD2, and SOD were all up-regulated in the radish root exposed to heat stress, respectively (Xu et al. 2006; Wang et al. 2018), indicating efficient ROS detoxification system in radish under stress.

Heat Stress Effects on Carbohydrate Metabolism Under HT, all glycolysis, citric-acid-related proteins were up-regulated, including glucose-6-phosphate isomerase, enolase, transaldolase, N-glyceraldehyde 3-phosphate dehydrogenase, citrate synthase, and aconitate hydratase (Shi et al. 2017). The enzymes like glucose-6-phosphate 1-dehydrogenase and ribulose-phosphate 3-epimerase involved in pentose phosphate pathway were inhibited by heat stress. Furthermore, other proteins such as galactose kinase, UDP-glucose 4-epimerase, V-type proton ATPase, and ATP/ADP (adenosine triphosphate/diphosphate) translocator were up-regulated, suggesting their potential role in plant stress response (Shi et al. 2017). Under heat stress carbohydrate metabolism has been thoroughly studied in many crop plants such as in onion (Galsurker et al. 2018), *Pyropia* (Shi et al. 2017), tangor (Moreno et al. 2018), barley (Hasanuzzaman et al. 2013), fescue (Xu et al. 2018), and rice (Islam et al. 2019).

It has also been reported in barley that the concentration of starch, fructose, and raffinose reduced under heat stress (Hasanuzzaman et al. 2013). Genes encoding enzymes involved in the synthesis of osmoprotectants and carbohydrate metabolism were up-regulated in the inner scales. Fructan is a polymer of fructose that serves as an important storage carbohydrate in onion. The fructan biosynthesis- elated gene *SST* was over-represented by the studies of Pollock and Cairns (1991) and Galsurker et al. (2018) in onion under different abiotic stress conditions (Hendry and Wallace 1993; Spollen and Nelson 1994; Galsurker et al. 2018) and it may have a role in protecting plants under stress. Turanose is an analogue of sucrose and a role as a signaling molecule was proposed by Sinha et al. (2002), who found that MAPK (mitogen-activated protein kinase) activity was strongly activated by turanose but not by sucrose, revealing different disaccharide specific pathways under heat stress (Moreno et al. 2018). Busch et al. (2005) suggest that heat stress induces the production of specific sugars, such as galactinol by galactinol synthases, and raffinose is involved in *Arabidopsis* heat-stress response. The accumulation of HSPs is a high energy consumption process in plants, and hence carbohydrate consumption increases (Finka et al. 2012; Mittler et al. 2012; Liu et al. 2013). Therefore, sugar metabolism or transport may play a key role in the process of ABA-induced HSPs accumulation in plants, as the adequate carbohydrates and their normal metabolism can provide energy to maintain higher HSPs accumulation and other physiological activities during the heat-stress period (Islam et al. 2019).

Heat Stress Effects on Lipid Metabolism Membrane plays an important role in response to environmental change, owing to selective permeability nature (Niu and Xiang 2018). Membrane is composed of lipid in bilayer structure, owing to its amphipathic nature. Lipids such as phosphatidylcholine (PC), glycolipids, sphingomyelin, phosphatidylserine (PS), phosphatidylethanolamine (PE), phosphatidylinositol (PI), phosphatidic acid, phosphatidylinositol-4-phosphate (PIP), and phosphatidylinositol-4-5 biphosphate (PIP_2) are selectively embedded in membrane bilayers (Andersen et al. 2016; Niu and Xiang 2018). Membranes are considered the primary site for injury under heat stress in plants (Nadeem et al. 2018).

Plants get adapted to temperature stress by modulating membrane fluidity which is affected by the degree of unsaturated fatty acid and changing in relative proportion of various lipid biomolecules throughout membrane bilayers (Li et al. 2016). Changes of glycerolipid composition in chloroplast are required for plants to accommodate heat stress response (Los et al. 2013; Boca et al. 2014; Routaboul et al. 2014; Chen et al. 2006). Several studies have focused on alteration in lipids under heat stress in different crop plants such as *Arabidopsis* (Routaboul et al. 2014; Higashi et al. 2015; Tang et al. 2016), tomato (Spicher et al. 2016), and wheat (Narayanan et al. 2016, 2018) to assess temperature stress tolerance potential. The trienoic fatty acid is a major component of thylakoid membrane, and thylakoid membrane is important for photosynthetic apparatus, which protects plants from heat-stress-related damage (Iba 2002; Wallis and Browse 2002). Routaboul et al. (2014) confirm the involvement of trienoic fatty acids in enhancing short- and long-term thermotolerance in *Arabidopsis* using triple mutant. To find out thermotolerance and better photosynthesis capacity in the *Arabidopsis* triple mutant *fad7-fad3-fad8* and wild-type plants transferred from 22 to 33 °C. After transferring to 33 °C, mutant plants showed slow growth rate, necrosis, damaged leaves, brown color plant at last plant died, while in the case of wild type of plants it remained healthy, continued to grow, susceptibility to heat stress. Photosynthetic rate Fv/Fm in wild type declined by less than 5%, but in mutant case it declined rapidly. These results indicated that the death of mutant plant is due to deficiency in trienoic fatty acid but it is present in wild-type plants. Also, it presents requirement of trienoic fatty acid for long-term thermotolerance in plants (Routaboul et al. 2014).

Higashi et al. (2015) determine the molecular regulation of lipid metabolism and stress responsive transcription factors (TFs) during heat stress in *Arabidopsis* (*fad3-2fad7-2fad8*) plants. Under 38 °C of heat stress, *Arabidopsis* plants showed induction of 60 species of lipids such as glycolipids monogalactosyldiacylglycerol (MGDG), phosphatidylglycerol (PG), PE, PI, diacylglycerol (DAG), and storage lipid triacyl glycerol (TAG), and it is decreased at the stress recovery phase. Out of all lipid species, triacyl glycerol species (54 : 9) containing α-linolenic acid (18 : 3) increased most under heat stress. Also many heat stress responsive TFs like *DREB2A*, *HSFA2* and *DREB1A* increase up to twofold, which showed the involvement of *Arabidopsis* plants in response to heat stress (Higashi et al. 2015).

Spicher et al. (2016) analyse changes in lipid composition in response to high (38 °C) and low temperature (10 °C and 20 °C) stress and its impact on the photosynthetic efficiency of tomato. After imposition of heat stress on tomato at 38 °C for six days, followed by five days of recovery, plants showed reduction in chlorophyll content; however, measured photochemical efficiency or quantum yield of PSII (*Fv/Fm*) values didn't show any significant effect on photosynthesis (Spicher et al. 2016). In contrast, 10 °C treatment exhibited photoinhibition and reduction of *Fv/Fm* was around 4%. Using lipidomic method, 791 species of lipids were detected – such as prenylquinones, plastoquinol (PQH2-9), plastoquinone (PQ-9), α-tocopherol, carotenoids, plastochromanol, hydroxyl-plastoquinol (PQ-OH), PC-8, hydroxyl plastochromanol (PC-OH) – and these were increased under high temperature and decreased level with high saturated fatty acids under low temperature. The study reveals that lipids which were highly expressed under heat stress protected PSII act as best electron carriers in the photosynthetic electron transport chain, scavenger of singlet oxygen species, inhibitor of lipid peroxidation, influence physical properties to membrane, degree of saturated fatty acids provide membrane fluidity; while low temperature stress presented photoinhibition, destabilization of biophysical properties, slow antioxidant biosynthesis, and insufficient lipid level to cope with this temperature. This investigation strongly suggests that an increased concentration of lipids is crucial for protection against high temperature stress and proper function of the photosynthetic apparatus (Spicher et al. 2016).

Temperature causes lipid alteration which is associated with heat tolerance and susceptibility in plants (Narayanan et al. 2015). Several studies specify that the heat stress increases the membrane damage and decreases antioxidant level to fight with ROS in wheat at the seedling, anthesis, and grain filling stages (Narayanan et al. 2018; Gupta and Mohan 2015). Narayanan et al. (2016) report that high day and night temperatures in wheat plant caused alteration in lipid metabolism and those lipids might have been associated with heat tolerance or susceptibility.

11.4.1.6 Heat Stress Effects on Secondary Metabolism

Under heat stress the genes involved in secondary metabolism, mainly in the phenylpropanoid- and flavonoid-biosynthesis pathways, were highly induced (Galsurker et al. 2018) and those includes genes encoding PAL1, C4H, 4CL, CHS, and F3H. PAL is the first dedicated enzyme in the phenylpropanoid pathway, converting phenylalanine into trans-cinnamic acid and providing a substrate for further synthesis of phenolic compounds (Hahlbrock and Scheel 1989; Galsurker et al. 2018). Several enzymes that participate in isoprenoid, terpenoid, riboflavin, and nucleotide sugars metabolism such as 4-hydroxy-3-methylbut-2-en-1-yl diphosphate synthase, farnesyltransferase, 1-deoxy-D-xylulose 5-phosphate reductoisomerase, flavin adenine dinucleotide (FAD) synthetase, and UDP-sulfoquinovose synthase decreased under heat stress (Shi et al. 2017). Genes encoding peroxidase (POD) and CYP450 were also highly represented in the outer scales in response to heat treatment. CYP450 is a large plant protein family that might contribute to extensive modifications of various secondary compounds and plays an important role in flavone biosynthesis and is expressed during tissue browning (Wang et al. 2013; Galsurker et al. 2018). The accretion of secondary metabolites, besides the accumulation of Pro, has previously been reported to maintain cellular homeostasis under HT stress (Mishra et al. 2017). The increased activity of phenylalanine ammonia-lyase (PAL), the key enzyme for secondary metabolites production, has been implicated under various abiotic stresses (Wahid and Ghazanfar 2006; Mishra et al. 2017).

11.4.2 Transcriptional Regulation

Genome-wide transcriptome analysis explained that there is some induction or repression of stress-responsive genes such as TFs, which play a vital role in response to various environmental stresses (Erpen et al. 2017) including heat stress (Ohama et al. 2017). To cope with various pattern of heat stress plants have HSF families including 10 members with different expression and function in the heat stress response and also get categorized according to their DNA-binding domains (Scarf et al. 2012). TFs are involved in various stress signaling pathways by binding to the *cis*-element in the promoter region of various stress responsive genes, which modulate a plant's adaptation to adverse conditions and enhance stress tolerance (Buscaill and Rivas 2014; Hernandez-Garcia and Finer 2014). So overexpression of these constitutively expressing HSFs, dehydration responsive element binding (DREB), WRKY, MYB (myeloblastosis) and basic leucine zipper (bZIP) (can induce many HSP genes and improved thermotolerance and survivability of crop plants such as wheat (Kumar et al. 2018; Geng et al. 2018; Zhao et al. 2017), *Arabidopsis* (Wu et al. 2018; He et al. 2016; Kidokoro et al. 2015), pepper (Dang et al. 2013), maize (Jiang et al. 2018), and rice (El-Kereamy et al. 2012).

In the heat acclimation process, HSFs play a central role in regulation of heat-inducible genes (Yabuta 2016; Fragkostefanakis et al. 2016; Liu and Chang 2013). Kumar et al. (2018) have studied heat-responsive TF (*TaHSFA6e*) genes in wheat cvs. Halna, HD2329, PBW343, and HD2985 under different heat stress conditions (30 and 38 °C, two-hour) at pollination and grain filling stage. Under heat stress, cv. Halna showed maximum relative expression of *TaHSFA6e* up to 3.6-fold as compared to other cultivars and controls. Overexpressing *TaHSFA6e* plants showed upregulation of stress-associated genes (SAG), such as *HSP70, HSP90, HSP17* under heat stress, which has remarkable effect on the thermotolerance of wheat at different stages of wheat growth. Wheat cv. Halna expressed better thermotolerant capacity by increasing total antioxidant capacity (TAC), catalase, guaiacol peroxidase compare to other cultivars (Kumar et al. 2018). Dehydration responsive element bindings (DREB2) TFs are reported to play a crucial role in heat stress and their overexpression can lead to improved thermotolerance of plants (Liu et al. 2014). Wu et al. (2018) report that a novel DREB2 gene (*LlDREB2B*) from lily and overexpressed in *Arabidopsis* to improve acquired thermotolerance under heat stress. After seven days of heat stress treatment up to 45 °C, over-expressed *LlDREB2B Arabidopsis* plants showed better growth with higher survival rates up to 100% than wild type plants. Transgenic *Arabidopsis* plants showed upregulation of some transcripts *AtHsp101, AtHsa32, AtHsp22.0, AtGolS1, AtABI5,* and *AtAPX2,* which might contribute stronger thermotolerance in *Arabidopsis* plants (Wu et al. 2018). CaWRKY40, a WRKY proteins found to be involved in pepper (*Capsicum annuum*) plants under HT and regulated thermotolerance (Dang et al. 2013).

Dang et al. (2013) have studied thermotolerance capacity of CaWRKY40 by over-expressing in tobacco plants and treated under heat stress of 42 °C for 48 hours. CaWRKY40-OE tobacco plants exhibited enhanced thermotolerance with higher survival rates of up to 90%. Transgenic tobacco plants up-regulated some heat-responsive-associated genes, ROS detoxified genes and three heat shock genes which were involved in heat tolerance of plants (Dang et al. 2013). However, studies are limited on MYB TF regarding heat stress. El-Kereamy et al. (2012) demonstrate that a rice R2-R3 MYB TFs OsMYB55 can enhance thermotolerance and improve grain yield and plant growth in *Oryza sativa* cv. Kaybonnet under HT stress (45 °C for 24 hours). Over-expression of OsMYB55 plants moderated great plant dry biomass, plant height and leaf sheath length under heat stress condition. Under adverse conditions, transgenic OsMYB55 plants expressed increase in amino acid metabolism by raising the level of gamma-aminobutyric acid (GABA), Pro, arginine, and glutamic acid, which leads to increased heat tolerance in rice plants (El-Kereamy et al. 2012). Geng et al. (2018) report that wheat *TabZIP60* TF confers heat tolerance in transgenic

Arabidopsis plants under heat stress treatment of 39 °C for one to three hours. Under heat stress *35S::TabZIP60s-MYC Arabidopsis* lines expressed 74.8% of seed germination and 48.2% of survival rate than in control plants (Geng et al. 2018). As noted above, recent studies on different TFs have revealed that the use of several heat stress-responsive TFs can contribute heat tolerance in crops plants.

11.4.3 Role of Small RNAs (miRNAs) in Heat-stress Tolerance

MicroRNAs are small noncoding RNAs of 19–25 nucleotides that play a vital role in heat stress responses in crop plants (Hasanuzzaman et al. 2013). Also miRNAs play a major role in plant growth and development (Liu et al. 2018), hormonal response (Curaba et al. 2014), stress defense such as drought (Wu et al. 2017), salinity (Lotfi et al. 2017), cold (Song et al. 2017), and oxidative (Iyer et al. 2012). A series of miRNAs were identified in regulating the response to heat stress in different plants such as rice (Mangrauthia et al. 2017), wheat (Kumar et al. 2014), laminariales (Liu et al. 2015), *Arabidopsis* (Stief et al. 2014), barley (Kruszka et al. 2014), and *Brassica rapa* (Yu et al. 2012). Under heat stress, miRNA regulates numerous stress-responsive genes (SAGs) which involve in various biological pathways for defense mechanism of plant to acquire thermotolerance (Fahlgren et al. 2007). In response to heat stress, up-regulated miRNA may down-regulate their specific genes in which miRNA act as negative regulators of thermotolerance and in opposite case down-regulated miRNA may up-regulate their target gene and influence thermotolerance (Nadeem et al. 2018).

Kruszka et al. (2014) reveal that induction of miRNA under heat stress in barley plants resulted in down-regulation of the expression level of novel target genes. Under heat stress of 35.5 °C for 6 and 24 hours, four mature miRNA were increased with their fold-change, i.e. miR166 (13-fold), miR167 (2.6-fold), miR160 (2.1-fold), and miR5175 (34-fold). Down-regulation of target gene occurred, such as *PHV* gene (auxiliary meristem initiation and leaf development), *homeobox leucine zipper protein HOX9-like* gene (embryogenesis), and *ARF13/17/8* gene (auxin responsive factor). Their study revealed that all four miRNA act as negative regulators of thermotolerance in barley plants (Kruszka et al. 2014). The overexpression of miRNA-resistant target genes plays an important role in plants through post-transcriptional gene silencing and may result in the overexpression of improved phenotypes (Nadeem et al. 2018).

Stief et al. (2014) define an miR156 in *Arabidopsis* which modulated the response to recurring heat stress through down-regulation of SPL gene and also modified growth and development after stress exposure. Under heat stress for 80–115 minutes in *Arabidopsis* plant, both involvement of AGO1 and DCL1 suggested that miRNA biogenesis is required for heat stress memory. Knockdown lines of *miR156a-f* (*Pro35S:MIM156h*) were deficient in the maintenance of both acquired and basal thermotolerance, also deficit in heat stress memory, reduce growth rate, early flowering with fewer leaves, and a reduced leaf initiation rate. Overexpression of *miR156a-f* lines (*Pro35S:MIR156h*) exhibited acquired thermotolerance, increased survival and growth rate, and very late flowering. *SPL2* and *SPL11* repression in miR156 lines showed heat stress memory-related genes, such as *APX2, HSA32, HSFA2, HSP17.6A, HSP22.0,* and *DIN2*. Overall data presented that miR156 is crucial for heat stress memory in plants (Stief et al. 2014).

Liu et al. (2015) identify a set of heat-stress-responsive miRNA and inspect their regulation during the heat stress response in *Saccharina japonica*, a temperate cold species when exposed to heat stress at 20 °C for three hours. A total of 75 novel miRNA were recognized in the control (CO) and heat stress libraries. Out of 75 novel miRNA, heat stress library expresses most up-regulated novel-mir-59 with a log-fold change of 15.99 and down-regulated novel mir-5 with a log-fold change of 13.43 compared to CO libraries. These novel miRNAs were involved in many pathways such as aminoacyl tRNA biosynthesis, ubiquitin-mediated proteolysis, protein processing in endoplasmic reticulum, protein and amino acid metabolism, carbon and carbohydrate metabolism, pyrimidine, purine, and RNA metabolism under heat stress. The study revealed that these miRNAs and its way of involvement in different pathways under heat stress played an important role of miRNAs in *S. japonica* heat-stress tolerance. This study also gives a foundation for future genetic improvement of *Saccharina* for heat-stress tolerance (Liu et al. 2015).

The identification of miRNA in wheat crop is very much less as compared to other crops (Kurtoglu et al. 2014). Kumar et al. (2014) report conserved and novel heat-responsive miRNA and their specific target in different tissues such as root, stem, and leaves of wheat cv. HD2985 under heat treatment of 42 °C for two hours. Several novel miRNAs were identified, such as candidate_430, candidate_3466, candidate_5064, candidate_5652, candidate_5993, candidate_3182, and candidate_6941. Out of seven miRNAs, candidate_3466 and candidate_5064 showed up-regulation and candidate_6941, 3182, 5652, 430 exhibited down-regulation under heat stress. These stress-responsive miRNAs are responsible for regulation of SAGs (their targets), which successively modulate the defense mechanism of wheat plant under heat stress. Eight conserved miRNAs were selected randomly under heat stress for the analysis of their targets, such as tae-miR156, tae-miR160, tae-miR159a, tae-miR167, tae-miR1117, tae-miR164, tae-miR319, and tae-miR398. It was found tae-miR156 (target *HSP90*), tae-miR160 (target *HSP70*), tae-miR159 (target *WRKY TF*), tae-miR1117 (target *CPK*), tae-miR164 (target *HSP17*), tae-miR398 (target *Cu/Zn-SODs*), and tae-miR167 (target *Dnaj*) showed up-regulation where only tae-miR319 exhibited maximum down-regulation in response to heat stress. The result concluded that up-regulation of both miRNAs and their targets suggest involvement of miRNAs in thermotolerance pathways (Kumar et al. 2014). Lin et al. (2018) studied the function of miR160 during heat stress at 44 °C for one hour in *Arabidopsis* plant by over-expressing miR160 precursor a (160OE) and artificial miR160 (MIM160), which resulted in plant development and survivability. According to the expression of miR160 under heat stress in both 160OE and MIMI160 plants, only 160OE exhibited higher regulation of miR160 compared to MIMI160. Owing to higher expression of miR160, 160OE plants enhanced seed germination rate of up to 80–94%, survival rate of up to 90%, and higher growth in hypocotyls and rachis length compared to MIMI160 and wild-type plants. Under heat stress overexpression of miR160 in 160OE plants enhanced the expression of *HSP21*, *HSP17.6A*, and *HSP17.6II*, which are involved in protection of plant cell from heat-induced programmed cell death, thermomemory, and enhanced heat tolerance. Also auxin response transcription factors (ARFs) expression negatively regulates thermotolerance, so miR160 could directly inhibit the function of ARF10, ARF17, and ARF16, which ultimately showed heat tolerance and heat sensitivity. Overall data presented that miR160 is an important regulator during heat stress in plants (Lin et al. 2018). Overall, miRNAs play a significant role in heat-stress tolerance in different crop plants.

11.5 Conclusion

Natural stress tolerance is a very complex process involving numerous metabolites and metabolic pathways. The thermoprotectants like HSPs, compatible solutes (such as GB, Pro, trehalose), hormones (like ABA, SA, etc.) and antioxidant system (such as CAT, SOD, and APX) make a major part of therm tolerance in crop plants under temperature stress. However, the signaling processes required for homeostasis of basic cellular and metabolic processes in adverse environments are just starting to emerge. Metabolic engineering of plants to synthesize compatible compounds may be an alternative way of developing thermotolerance in important crop plants. It necessitates researches in conducting studies on stress combinations (for example drought and HT) to decipher underlying molecular mechanism of plants to cope with HT stress, and such studies are still in their infancy. In addition, several transcriptome studies demonstrate that unique mechanisms that govern the responses of plants to stress combinations could not be easily predicted from the studies focusing on individual stress factors. Indeed, mechanisms that regulate defense responses in plants were shown to be different under different temperatures.

The application of newer technologies such as synthetic biology, gene pyramiding, and genome-editing tools are helpful in engineering HT stress defense pathways further and to enhance HT stress tolerance in crop plants. In addition, high throughput global "omics" approaches have a promising role in exploring gene function in connection with thermoprotectant

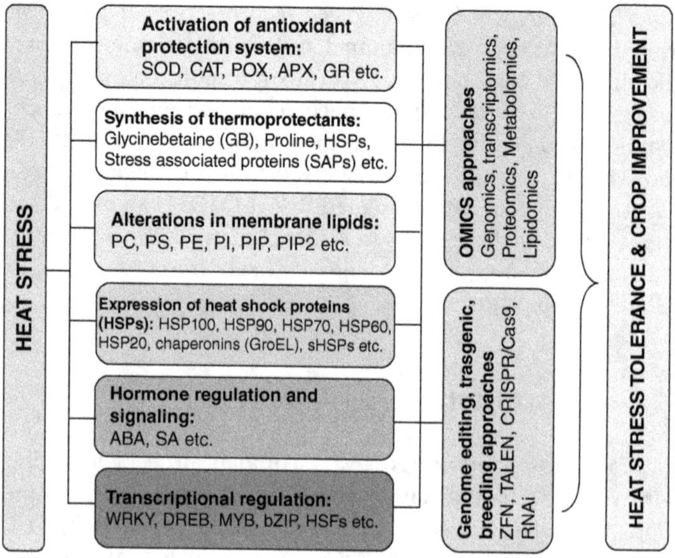

Figure 11.1 Heat stress defence responses, combined global approaches for identifying and validating candidate genes/traits for improving heat-stress tolerance in crop plants.

biosynthetic pathways, role of HSPs, antioxidative pathways, hormones and signaling networks, and crop HT stress responses, in a more holistic way (Figure 11.1). The ultimate practical goal of all these studies is to enhance temperature stress tolerance of crop plants, which can grow better under adverse environmental stress conditions worldwide. This study gives an overall and updated idea of plants' in-built defense mechanisms toward HT.

Acknowledgement

Work in the laboratory of GKS is supported by the Forest and Environment Department, Government of Odisha, India, and is gratefully acknowledged. The authors apologize for being unable to cite all relevant papers.

References

Ahanger, M.A., Akram, N.A., Ashraf, M. et al. (2017). Plant responses to environmental stresses from gene to biotechnology. *AoB Plants* 9: plx025. https://doi.org/10.1093/aobpla/plx025.

Ahmad, A., Diwan, H., and Abrol, Y.P. (2010). Global climate change, stress and plant productivity. In: *Abiotic Stressa daptation in Plants: Physiological, Molecular and Genomefoundation* (eds. A. Pareek, S.K. Sopory, H.J. Bohnert and Govindjee), 503–521. Dordrecht, the Netherlands: Springer Science +Business Media BV.

Ahmed, J.U. and Hassan, M.A. (2011). Evaluation of seedling proline content of wheat genotypes in relation to heat tolerance. *Bangladesh Journal of Botany* 40 (1): 17–22.

Ahmed, F.E., Hall, A.E., and Madore, M.A. (1993). Interactive effects of high temperature and elevated carbon dioxide concentration on cowpea (*Vigna unguiculata* [L.] Walp.). *Plant, Cell & Environment* 16: 835–842.

Alamri, S.A., Siddiqui, M.H., Al-Khaishany, M.Y. et al. (2018). Nitric oxide-mediated cross-talk of proline and heat shock proteins induce thermo tolerance in *Vicia faba* L. *Environmental and Experimental Botany* https://doi.org/10.1016/j.envexpbot.2018.06.012.

Allakhverdiev, S.I., Los, D.A., Mohanty, P. et al. (2007). Glycinebetaine alleviates the inhibitory effect of moderate heat stress on the repair of photosystem II during photoinhibition. *Biochimica et Biophysica Acta* 1767: 1363–1371.

Almeselmani, M., Deshmukh, P.S., Sairam, R.K. et al. (2006). Protective role of antioxidant enzymes under high temperature stress. *Plant science* 171 (3): 382–388.

Almeselmani, M., Deshmukh, P.S., and Sairam, R.K. (2009). High temperature stress tolerance in wheat genotypes: role of antioxidant defence enzymes. *Acta Agronomica Hungarica* 57: 1–14. https://doi.org/10.1556/AAgr.57.2009.1.1.

Andersen, J.P., Vestergaard, A.L., Mikkelsen, S.A. et al. (2016). P4-ATPases as phospholipid flippases: structure, function, and enigmas. *Frontiers in Physiology* 7: 275.

Antunes, S.C., Freitas, R., Figueira, E. et al. (2013). Biochemical effects of acetaminophen in aquatic species: edible clams *Venerupis decussata* and *Venerupis philippinarum*. *Environmental Science and Pollution Research International* 20 (9): 6658–6666. https://doi.org/10.1007/s11356-013-1784-9.

Ashraf, M. and Foolad, M.R. (2007). Roles of glycine betaine and proline in improving plant abiotic stress resistance. *Environmental and Experimental Botany* 59: 206–216.

Balfagón, D., Zandalinas, S.I., and Gómez-Cadenas, A. (2018). High temperatures change the perspective: integrating hormonal responses in citrus plants under co-occurring abiotic stress conditions. *Physiologia Plantarum* https://doi.org/10.1111/ppl.12815.

Barghi, S.S., Mostafaii, H., Peighami, F., and Zakaria, R.A. (2012). Path analysis of yield and its components in lentil under end season heat condition. *International Journal of Agricultural Research and Reviews* 2: 969–974.

Board, J.E. and Kahlon, C.S. (2011). Soybean yield formation: what controls it and how it can be improved. *Soybean Physiology and Biochemistry*: 1–36.

Boca, S., Koestler, F., Ksas, B. et al. (2014). *Arabidopsis* lipocalins AtCHL and AtTIL have distinct but overlapping functions essential for lipid protection and seed longevity. *Plant, Cell & Environment* 37 (2): 368–381.

Buscaill, P. and Rivas, S. (2014). Transcriptional control of plant defence responses. *Current Opinion in Plant Biology* 20: 35–46.

Busch, W., Wunderlich, M., and Schoffl, F. (2005). Identification of novel heat shock factor-dependent genes and biochemical pathways in *Arabidopsis thaliana*. *The Plant Journal* 41 (1): 1–14.

Chavez-Arias, C.C., Ligarreto-Moreno, G.A., and Restrepo-D'ıaz, H. (2018). Evaluation of heat stress period duration and the interaction of daytime temperature and cultivar on common bean. *Environmental and Experimental Botany* https://doi.org/10.1016/j.envexpbot.2018.08.012.

Chen, T.H. and Murata, N. (2011). Glycinebetaine protects plants against abiotic stress: mechanisms and biotechnological applications. *Plant, Cell & Environment* 34: 1–20.

Chen, J., Burke, J.J., Xin, Z. et al. (2006). Characterization of the Arabidopsis thermosensitive mutant atts02 reveals an important role for galactolipids in thermotolerance. *Plant, Cell & Environment* 29: 1437–1448.

Chen, J., Wang, P., Mi, H.L. et al. (2011). Reversible association of ribulose-1, 5-bisphosphate carboxylase/oxygenase activase with the thylakoid membrane depends upon the ATP level and pH in rice without heat stress. *Journal of Experimental Botany* 61: 2939–2950. https://doi.org/10.1093/jxb/erq122.

Cheng, T., Shi, J., Dong, Y. et al. (2017). Hydrogen sulfide enhances poplar tolerance to high-temperature stress by increasing *S*-nitrosoglutathione reductase (GSNOR) activity and reducing reactive oxygen/nitrogen damage. *Plant Growth Regulation* https://doi.org/10.1007/s10725-017-0316-x.

Chhabra M.L, Dhawan A, Sangwan N, Dhawan K, Singh D (2009). Phytohormones induced amelioration of high temperature stress in *Brassica juncea* (L.) Czern & Coss. Proceedings of 16th Australian Research Assembly on Brassicas, 10th–14th September Ballarat, Australia.

Cleland, D., Krader, P., McCree, C. et al. (2004). Glycine betaine as a cryoprotectant for prokaryotes. *Journal of Microbiological Methods* 58: 31–38.

Coqueiro, D.S., de Souza, A.A., Takita, M.A. et al. (2015). Transcriptional profile of sweet orange in response to chitosan and salicylic acid. *BMC Genomics* 16: 288. https://doi.org/10.1186/s12864-015-1440-5.

Curaba, J., Singh, M.B., and Bhalla, P.L. (2014). miRNAs in the crosstalk between phytohormone signalling pathways. *Journal of Experimental Botany* 65: 1425–1438.

Cvikrová, M., Gemperlová, L., Dobrá, J. et al. (2012). Effect of heat stress on polyamine metabolism in proline-over-producing tobacco plants. *Plant Science* 182: 49–58.

Dang, F.F., Wang, Y.N., Yu, L. et al. (2013). CaWRKY40, a WRKY protein of pepper, plays an important role in the regulation of tolerance to heat stress and resistance to *Ralstonia solanacearum* infection. *Plant, Cell & Environment* 36: 757–774.

Das, S., Krishnan, P., Nayak, M., and Ramakrishnan, B. (2014). High temperature stress effects on pollens of rice (*Oryza sativa* L.) genotypes. *Environmental and Experimental Botany* 101: 36–46.

Dat, J.F., Foyer, C.H., and Scott, I.M. (1998). Changes in salicylic acid and antioxidants during induced thermotolerance in mustard seedlings. *Plant Physiology* 118 (4): 1455–1461.

Dempsey, D.A., Vlot, A.C., Wildermuth, M.C., and Klessig, D.F. (2011). Salicylic acid biosynthesis and metabolism. *Arabidopsis Book* 9: e0156. https://doi.org/10.1199/tab.0156.

Devasirvatham, V., Tan, D.K.Y., Gaur, P.M. et al. (2012). High temperature tolerance in chickpea and its implications for plant improvement. *Crop & Pasture Science* 63: 419–428. https://doi.org/10.1071/CP11218.

Devasirvatham, V., Gaur, P.M., Mallikarjuna, N. et al. (2013). Reproductive biology of chickpea response to heat stress in the field is associated with the performance in controlled environments. *Field Crops Research* 142: 9–19. https://doi.org/10.1016/j.fcr.2012.11.011.

Ding, H., He, J., Wu, Y. et al. (2018). The tomato mitogen-activated protein kinase SlMPK1 is as a negative regulator of the high temperature stress response. *Plant Physiology* https://doi.org/10.1104/pp.18.00067.

Djanaguiraman, M., Prasad, P.V.V., and Seppanen, M. (2010). Selenium protects sorghum leaves from oxidative damage under high temperature stress by enhancing antioxidant defense system. *Plant Physiology and Biochemistry* 48: 999–1007.

Djanaguiraman, M., Perumal, R., Jagadish, S.V.K. et al. (2017). Sensitivity of sorghum pollen and pistil to high-temperature stress. *Plant, Cell & Environment* 41 (5): 1065–1082. https://doi.org/10.1111/pce.13089.

Djanaguiraman, M., Boyle, D.L., Welti, R. et al. (2018). Decreased photosynthetic rate under high temperature in wheat is due to lipid desaturation, oxidation, acylation, and damage of organelles. *BMC Plant Biology* 18: 55. https://doi.org/10.1186/s12870-018-1263-z.

Driedonks, N., Rieu, I., and Vriezen, W.H. (2016). Breeding for plant heat tolerance at vegetative and reproductive stages. *Plant Reproduction* 29 (1–2): 67–79. https://doi.org/10.1007/s00497-016-0275-9.

El-kereamy, A., Bi, Y.M., Ranathunge, K. et al. (2012). The rice R2R3-MYB transcription factor OsMYB55 is involved in the tolerance to high temperature and modulates amino acid metabolism. *PLoS One* 7: 1–16.

Erpen, L., Sunitibala Devi, H., Grosser, J.W., and Dutt, M. (2017). Potential use of the DREB/ERF, MYB, NAC and WRKY transcription factors to improve abiotic and biotic stress in transgenic plants. *Plant Cell, Tissue and Organ Culture* 132: 1–25.

Fahlgren, N., Howell, M.D., Kasschau, K.D. et al. (2007). High throughput sequencing of Arabidopsis microRNAs: evidence for frequent birth and death of MIRNA genes. *PLoS One* 2 (2): e219.

Farag, M.A., Meyer, A., Ali, S.E. et al. (2018). A comparative metabolomics approach detects stress specific responses during coral bleaching in soft corals. *Journal of Proteome Research* https://doi.org/10.1021/acs.jproteome.7b00929.

Fariduddin, Q., Varshney, P., Yusuf, M. et al. (2013). Dissecting the role of glycinebetaine in plants under abiotic stress. *Plant Stress*: 1–18.

Feller, U. (2016). Drought stress and carbon assimilation in a warming climate: reversible and irreversible impacts. *Journal of Plant Physiology* https://doi.org/10.1016/j.jplph.2016.04.002.

Finka, A., Cuendet, A.F.H., Maathuis, F.J. et al. (2012). Plasma membrane cyclic nucleotide gated calcium channels control land plant thermal sensing and acquired thermotolerance. *Plant Cell* 24 (8): 3333–3348.

Finkelstein, R. (2013). Abscisic acid synthesis and response. *The Arabidopsis Book* 11: e0166.

Fragkostefanakis, S., Mesihovic, A., Simm, S. et al. (2016). HsfA2 controls the activity of developmentally and stress-regulated heat stress protection mechanisms in tomato male reproductive tissues. *Plant Physiology* 170 (4): 2461–2477.

Frank, G., Pressman, E., Ophir, R. et al. (2009). Transcriptional profiling of maturing tomato (*Solanum lycopersicum* L.) microspores reveals the involvement of heat shock proteins, ROS scavengers, hormones, and sugars in the heat stress response. *Journal of Experimental Botany* 60 (13): 3891–3908.

Galsurker, O., Doron-Faigenboim, A., Teper-Bamnolker, P. et al. (2018). Differential response to heat stress in outer and inner onion bulb scales. *Journal of Experimental Botany* 69 (16): 4047–4064. https://doi.org/10.1093/jxb/ery189.

Gan Y, Wang Y, Angadi S.V, Mcdonald C. L (2004). Response of chickpea to short periods of high temperature and water stress at different developmental stages. *4th International Crop Science Congress*, Brisbane, Australia 26th September to 1st October.

Geng, X., Zanga, X., Lia, H. et al. (2018). Unconventional splicing of wheat TabZIP60 confers heat tolerance in transgenic *Arabidopsis*. *Plant Science* 274: 252–260.

Giorno, F., Wolters-Arts, M., Mariani, C., and Rieu, I. (2013). Ensuring reproduction at high temperatures: the heat stress response during anther and pollen development. *Plants* 2: 489–506. https://doi.org/10.3390/plants2030489.

Giri, J. (2011). Glycinebetaine and abiotic stress tolerance in plants. *Plant Signaling & Behavior* 6: 1746–1751.

Gómez-Cadenas, A., Manzi, M., Lado, J. et al. (2015). Root ABA accumulation in long-term water-stressed plants is sustained by hormone transport from aerial organs. *Plant & Cell Physiology* 56 (12): 2457–2466.

Gong, M., Li, Y.J., and Chen, S.Z. (1998). Abscisic acid-induced thermotolerance in maize seedlings is mediated by calcium and associated with antioxidant systems. *Journal of Plant Physiology* 153 (3–4): 488–496.

Green, A., Friebe, B., Prasad, P.V.V., and Fritz, A. (2018). Evaluating heat tolerance of a complete set of wheat-*Aegilops geniculata* chromosome addition lines. *Journal of Agronomy and Crop Science* 2018: 1–6. https://doi.org/10.1111/jac.12282.

Gupta, R.K. and Mohan, D. (2015). Relevance of physiological efficiency in wheat grain quality and the prospects of improvement. *Physiology and Molecular Biology of Plants* 21 (4): 591–596.

Hahlbrock, K. and Scheel, D. (1989). Physiology and molecular biology of phenylpropanoid metabolism. *Annual Review of Plant Biology* 40 (1): 347–369.

Hall, A.E. (1992). Breeding for heat tolerance. *Plant Breeding Reviews* 10: 129–168.

Harsh, A., Sharma, Y.K., Joshi, U. et al. (2016). Effect of short-term heat stress on total sugars, proline and some antioxidant enzymes in moth bean (*Vigna aconitifolia*). *Annals of Agricultural Science* 61: 57–64. https://doi.org/10.1016/j.aoas.2016.02.001.

Hasanuzzaman, M., Nahar, K., Alam, M. et al. (2013). Physiological, biochemical, and molecular mechanisms of heat stress tolerance in plants. *International Journal of Molecular Sciences* 14 (5): 9643–9684. https://doi.org/10.3390/ijms14059643.

Hatfield, J.L., Boote, K.J., Kimball, B.A. et al. (2011). Climate impacts on agriculture: implications for crop production. *Agronomy Journal* 103: 351–370. https://doi.org/10.2134/agronj2010.0303.

He, G.-H., Xu, J.-Y., Wang, Y.-X. et al. (2016). Drought-responsive WRKY transcription factor genes *TaWRKY1* and *TaWRKY33* from wheat confer drought and/or heat resistance in *Arabidopsis*. *BMC Plant Biology* 16: 1–16.

Hendry, G.A. and Wallace, R.K. (1993). The origin, distribution, and evolutionary significance of fructans. *Science and Technology of Fructans*: 119–139.

Hernandez-Garcia, C.M. and Finer, J.J. (2014). Identification and validation of promoters and *cis*-acting regulatory elements. *Plant Science* 217–218: 109–119.

Higashi, Y., Okazaki, Y., Myouga, F. et al. (2015). Landscape of the lipidome and transcriptome under heat stress in Arabidopsis thaliana. *Scientific Reports* 5: 10533.

Higashitani, A. (2013). High temperature injury and auxin biosynthesis in micro sporogenesis. *Frontiers in Plant Science* 4: 47. https://doi.org/10.3389/fpls.2013.00047.

Hirayama, T. and Shinozaki, K. (2010). Research on plant abiotic stress responses in the post-genome era: past, present and future. *The Plant Journal* 61 (6): 1041–1052.

Iba, K. (2002). Acclimative response to temperature stress in higher plants: approaches of gene engineering for temperature tolerance. *Annual Review of Plant Biology* 53: 225–245. https://doi.org/10.1146/annurev.arplant.53.100201.160729.

Intergovernmental Panel on Climate Change (IPCC) (2014). *Climate Change 2014: Impact Adaptation and Vulnerability*, 151. IPCC: Geneva, Switzerland.

Islam, M.T. (2011). Effect of temperature on photosynthesis, yield attributes and yield of aromatic rice genotypes. *International Journal of Sustainable Crop Production* 6: 14–16.

Islam, M.O., Kato, H., Shima, S. et al. (2019). Functional identification of a rice trehalase gene involved in salt stress tolerance. *Gene* 685: 42–49.

Iyer, N.J., Jia, X., Sunkar, R. et al. (2012). MicroRNAs responsive to ozone-induced oxidative stress in *Arabidopsis thaliana*. *Plant Signaling & Behavior* 7: 484–491.

Jiang, J., Liu, X., Liu, C. et al. (2017). Integrating omics and alternative splicing reveals insights into grape response to high temperature. *Plant Physiology* 173 (2): 1502–1518. https://doi.org/10.1104/pp.16.01305.

Jiang, Y., Zheng, Q., Chen, L. et al. (2018). Ectopic overexpression of maize heat shock transcription factor gene *ZmHsf04* confers increased thermo and salt-stress tolerance in transgenic *Arabidopsis*. *Acta Physiologiae Plantarum* 40 (9): 1–12.

Johkan, M., Oda, M., Maruo, T., and Shinohara, Y. (2011). Crop production and global warming. In: *Global Warming Impacts: Case Studies on the Economy, Human Health, and on Urban and Natural Environments* (ed. S. Casalegno), 139–152. Rijeka, Croatia: In Tech.

Karim, M.A., Fracheboud, Y., and Stamp, P. (1997). Heat tolerance of maize with reference of some physiological characteristics. *Annals of Bangladesh Agriculture* 7: 27–33.

Kaushal, N., Awasthi, R., Gupta, K. et al. (2013). Heat-stress-induced reproductive failures in chickpea (*Cicer arietinum*) are associated with impaired sucrose metabolism in leaves and anthers. *Functional Plant Biology* 40: 1334–1349. https://doi.org/10.1071/FP13082.

Kaushal, N., Bhandari, K., Siddique, K.H.M., and Nayyar, H. (2016, 2016). Food crops face rising temperatures: an overview of responses, adaptive mechanisms, and approaches to improve heat tolerance. *Cogent Food & Agriculture* 2: 1134380. https://doi.org/10.1080/23311932.2015.1134380.

Khodarahmpour, Z. (2011). Genetic analysis of tolerance to heat stress in maize (*Zea mays* L.). *African Journal of Agricultural Research* 6: 2767–2773.

Kidokoro, S., Watanabe, S., Ohori, T. et al. (2015). Soybean $DREB_1/CBF$-type transcription factors function in heat and drought as well as cold stress-responsive gene expression. *The Plant Journal* 81: 505–518.

Kim, K. and Portis, A.R. (2005). Temperature dependence of photosynthesis in *Arabidopsis* plants with modifications in Rubisco activase and membrane fluidity. *Plant & Cell Physiology* 46: 522–530. https://doi.org/10.1093/pcp/pci052.

Kotak, S., Larkindale, J., Lee, U. et al. (2007). Complexity of the heat stress response in plants. *Current Opinion in Plant Biology* 10: 310–316.

Krasensky, J. and Jonak, C. (2012). Drought, salt, and temperature stress-induced metabolic rearrangements and regulatory networks. *Journal of Experimental Botany* 63: 1593–1608.

Kruszka, K., Pacak, A., Swida-Barteczka, A. et al. (2014). Transcriptionally and post-transcriptionally regulated microRNAs in heat stress response in barley. *Journal of Experimental Botany* 65 (20): 6123–6135.

Kumar, S., Kaur, R., Kaur, N. et al. (2011). Heat-stress induced inhibition in growth and chlorosis in mungbean (*Phaseolus aureus* Roxb.) is partly mitigated by ascorbic acid application and is related to reduction in oxidative stress. *Acta Physiologiae Plantarum* 33: 2091–2101. https://doi.org/10.1007/s11738-011-0748-2.

Kumar, R.R., Goswami, S., Sharma, S.K. et al. (2012). Protection against heat stress in wheat involves change in cell membrane stability, antioxidant enzymes, osmolyte, H_2O_2 and transcript of heat shock protein. *International Journal of Plant Physiology and Biochemistry* 4 (4): 83–91.

Kumar, S., Thakur, P., Kaushal, N. et al. (2013). Effect of varying high temperatures during reproductive growth on reproductive function, oxidative stress and seed yield in chickpea genotypes differing in heat sensitivity. *Archives of Agronomy and Soil Science* 59: 823–843.

Kumar, M., Lee, S.C., and Kim, J.Y. (2014). Over-expression of dehydrin gene, OsDhn1, improves drought and salt stress tolerance through scavenging of reactive oxygen species in rice (*Oryza sativa* L.). *Journal of Plant Biology* 57: 383–393.

Kumar, R.R., Goswami, S., Singh, K. et al. (2018). Characterization of novel heat-responsive transcription factor (*TaHSFA6e*) gene in regulation of heat shock proteins (HSPs): a key member of heat stress tolerance network of wheat. *Journal of Biotechnology* 279: 1–12.

Kurtoglu, K.Y., Kantar, M., and Budak, H. (2014). New wheat microRNA using whole-genome sequence. *Functional & Integrative Genomics* 14: 363–379. https://doi.org/10.1007/s10142-013-0357-9.

Laghri, K.A., Mahboob, A.S., and Arain, M.A. (2012). Effect of high temperature stress on grain yield and yield components of wheat (*Triticum aestivum* L.). *Science, Technology and Development* 31: 83–90.

Larkindale, J. and Knight, M.R. (2002). Protection against heat stress-induced oxidative damage in *Arabidopsis* involves calcium, abscisic acid, ethylene, and salicylic acid. *Plant Physiology* 128: 682–695. https://doi.org/10.1104/pp.010320.

Larkindale, J. and Vierling, E. (2008). Core genome responses involved in acclimation to high temperature. *Plant Physiology* 146: 748–761.

Lee, C.B., Hayashi, H., and Moon, B.Y. (1997). Stabilization by glycinebetaine of photosynthetic oxygen evolution by thylakoid membranes from *Synechococcus* PCC7002. *Molecules and Cells* 7: 296–299.

Li, N., Xu, C., Li-Beisson, Y., and Philippar, K. (2016). Fatty acid and lipid transport in plant cells. *Trends in Plant Science* 21 (2): 145–158.

Lin, J.S., Kuo, C.C., Yang, I.C. et al. (2018). MicroRNA160 modulates plant development and heat shock protein gene expression to mediate heat tolerance in *Arabidopsis. Frontiers in Plant Science* 9: 68.

Liu, H.C. and Charng, Y.Y. (2013). Common and distinct functions of Arabidopsis class A1 and A2 heat shock factors in diverse abiotic stress responses and development. *Plant Physiology* 163: 276–290. https://doi.org/10.1104/pp.113.221168.

Liu, X., Wu, H., Ji, C. et al. (2013). An integrated proteomic and metabolomic study on the chronic effects of mercury in Suaeda salsa under an environmentally relevant salinity. *PLoS One* 8: e64041.

Liu, X., Zhang, H., Zhao, Y., Feng, Z., Li, Q., Yang, H.Q. (2014). Auxin controls seed dormancy through stimulation of abscisic acid signaling by inducing ARF-mediated ABI3 activation in Arabidopsis. Proceedings of the National Academy of Sciences USA, 110, 15485–15490. doi: https://doi.org/10.1073/pnas.13046\ignorespaces51110.

Liu, F., Wang, W., Sun, X. et al. (2015). Conserved and novel heat stress-responsive micro RNAs were identified by deep sequencing in *Saccharina japonica* (Laminariales, Phaeophyta). *Plant, Cell & Environment* 38 (7): 1357–1367.

Liu, H., Yu, H., Tang, G., and Huang, T. (2018). Small but powerful: function of microRNAs in plant development. *Plant Cell Reports* 37: 515–528.

Los, D.A., Mironov, K.S., and Allakhverdiev, S.I. (2013). Regulatory role of membrane fluidity in gene expression and physiological functions. *Photosynthesis Research* 116 (2–3): 489–509.

Lotfi, A., Pervaiz, T., Jiu, S. et al. (2017). Role of microRNAs and their target genes in salinity response in plants. *Plant Growth Regulation* 82: 377–390.

Maçãs, B., Gomes, M.C., Dias, A.S., and Coutinho, J. (2000). The tolerance of durum wheat to high temperatures during grain filling. In: *Durum Wheat improvement in the Mediterranean Region: New Challenges* (eds. C. Royo, M.M. Nachit, N. Di Fonzo and J.L. Araus), 257–261. Zaragoza, Spain: CIHEAM.

Mangrauthia, S.K., Bhogireddy, S., Agarwal, S. et al. (2017). Genome-wide changes in microRNA expression during short and prolonged heat stress and recovery in contrasting rice cultivars. *Journal of Experimental Botany* 68 (9): 2399–2412.

Martinez, V., Cordones, M.N., Delacalle, M.L. et al. (2018). Tolerance to stress combination in tomato plants: new insights in the protective role of melatonin. *Molecules* 2018 (23): 535. https://doi.org/10.3390/molecules23030535.

Mathur, S., Agrawal, D., and Jajoo, A. (2014). Photosynthesis: response to high temperature stress. *Journal of Photochemistry and Photobiology, B: Biology* 137: 116–126. https://doi.org/10.1016/j.jphotobiol.2014.01.010.

McNeil, S.D., Nuccio, M.L., and Hanson, A.D. (1999). Betaines and related osmo-protectants targets for metabolic engineering of stress resistance. *Plant Physiology* 120: 945–949.

Mingpeng, H., Yongge, G., Chengzhang, W. et al. (2010). Related studies on the effects of high temperature stress on alfalfa and its heat resistance mechanism. *Genomics and Applied Biology* 29: 563–569.

Mishra, D., Shekhar, S., Agrawal, L. et al. (2017, 2017). Cultivar-specific high temperature stress responses in bread wheat (*Triticum aestivum* L.) associated with physicochemical traits and defense pathways. *Food Chemistry* 221: 1077–1087. https://doi.org/10.1016/j.foodchem.2016.11.053.

Mittler, R., Finka, A., and Goloubinoff, P. (2012). How do plants feel the heat? *Trends in Biochemical Sciences* 37: 118–125.

Moreno, A.S., Perotti, V.E., Margarit, E. et al. (2018). Metabolic profiling and quality assessment during the postharvest of two tangor varieties subjected to heat treatments. *Postharvest Biology and Technology* 142: 10–18. https://doi.org/10.1016/j.postharvbio.2018.03.014.

Murata, N., Mohanty, P.S., Hayashi, H., and Papageorgiou, G.C. (1992). Glycinebetaine stabilizes the association of extrinsic proteins with the photosynthetic oxygen-evolving complex. *FEBS Letters* 296 (2): 187–189.

Murmu, K., Murmu, S., Kundu, C.K., and Bera, P.S. (2017). Exogenous proline and glycine betaine in plants under stress tolerance. *International Journal of Current Microbiology and Applied Sciences* 6 (9): 901–913. https://doi.org/10.20546/ijcmas.2017.609.109.

Nadeem, M., Li, J., Wang, M. et al. (2018). Unraveling field crops sensitivity to heat stress: mechanisms, approaches, and future prospects. *Agronomy* 8 (7): 128.

Narayanan, S., Prasad, P.V.V., Fritz, A.K. et al. (2015). Impact of high night-time and high daytime temperature stress on winter wheat. *Journal of Agronomy and Crop Science* 201: 206–218.

Narayanan, S., Tamura, P.J., Roth, M.R. et al. (2016). Wheat leaf lipids during heat stress: I: high day and night temperatures results in major lipid alternations. *Plant, Cell & Environment* 39: 787–803.

Narayanan, S., Prasad, P.V.V., and Welti, R. (2018). Alterations in wheat pollen lipidome during high day and night temperature stress. *Plant, Cell & Environment* 41 (8): 1749–1761. https://doi.org/10.1111/pce.13156.

Nayyar, H., Bains, T., and Kumar, S. (2005a). Low temperature induced floral abortion in chickpea: relationship to abscisic acid and cryoprotectants in reproductive organs. *Environmental and Experimental Botany* 53: 39–47. https://doi.org/10.1016/j.envexpbot.2004.02.011.

Nayyar, H., Chander, K., Kumar, S., and Bhains, T. (2005b). Glycinebetaine mitigates cold stress damage in chickpea. *Agronomy for Sustainable Development* 25: 381–388.

Niu, Y. and Xiang, Y. (2018). An overview of biomembrane functions in plant responses to high-temperature stress. *Frontiers in Plant Science* 9: 915.

Ohama, N., Sato, H., Shinozaki, K., and Shinozak, I.K.Y. (2016). Transcriptional regulatory network of plant heat stress response. *Trends in Plant Science* 1474 https://doi.org/10.1016/j.tplants.2016.08.015.

Ohama, N., Sato, H., Shinozaki, K., and Yamaguchi-Shinozaki, K. (2017). Transcriptional regulatory network of plant heat stress responses. *Trends in Plant Science* 22: 53–65.

Ortiz, C. and Cardemil, L. (2001). Heat-shock responses in two leguminous plants: a comparative study. *Journal of Experimental Botany* 52: 1711–1719.

Pan, Q., Zhan, J., Liu, H. et al. (2006). Salicylic acid synthesized by benzoic acid 2-hydroxylase participates in the development of thermotolerance in pea plants. *Plant Science* 171 (2): 226–233.

Papageogiou, G.C. and Murata, N. (1995). The unusually strong stabilizing effects of glycinebetaine on the structure and function of the oxygen-evolving photosystem II complex. *Photosynthesis Research* 44: 243–252.

Penfield, S. (2008). Temperature perception and signal transduction in plants. *The New Phytologist* 179 (3): 615–628.

Peters, G.P., Marland, G., Le Quere, C. et al. (2011). Rapid growth in CO_2 emissions after the 2008–2009 global financial crisis. *Nature Climate Change* 2: 2–4.

Piramila, B.H.M., Prabha, A.L., Nandagopalan, V., and Stanley, A.L. (2012). Effect of heat treatment on germination, seedling growth and some biochemical parameters of dry seeds of black gram. *International Journal of Pharmaceutical and Phytopharmacological Research* 1: 194–202.

Pollock, C.J. and Cairns, A.J. (1991). Fructan metabolism in grasses and cereals. *Annual Review of Plant Biology* 42 (1): 77–101.

Prasad, N.G., Dey, S., Shakarad, M., and Joshi, A. (2003). The evolution of population stability as a by-product of life history evolution. *Proceedings of the Royal Society B: Biological Sciences* 270: S84–S86. https://doi.org/10.1098/rsbl.2003.0020.

Prasad, P.V.V., Staggenborg, S.A., and Ristic, Z. (2008). Impacts of drought and/or heat stress on physiological, developmental, growth, and yield processes of crop plants. In: *Response of Crops to Limited Water: Understanding and Modeling Water Stress Effects on Plant Growth Processes*, Advances in Agricultural Systems Modeling Series 1 (eds. L.H. Ahuja and S.A. Saseendran), 301–355. Madison, WI: American Society of

Agronomy, Crop Science Society of America, Soil Science Society of America.

Prasad, P.V.V., Pisipati, S.R., Momcilovic, I., and Ristic, Z. (2011). Independent and combined effects of high temperature and drought stress during grain filling on plant yield and chloroplast EF-TU expression in spring wheat. *Journal of Agronomy and Crop Science* 197: 430–441.

Prasad, P.V.V., Bheemanahalli, R., and Jagadish, S.V.K. (2017). Field crops and the fear or heat stress: opportunities, challenges and future directions. *Field Crops Research* 200: 114–121.

Prasad, P.V.V., Djanaguiraman, M., Jagadish, S.V.K. et al. (2018). Drought and high temperature stress and traits associated with tolerance. *Agronomy Monographs* https://doi.org/10.2134/agronmonogr58.2014.0065.

Queitsch, C., Hong, S.W., Vierling, E., and Lindquest, S. (2000). Heat shock protein 101 plays a crucial role in thermotolerance in *Arabidopsis*. *Plant Cell Online* 12: 479–492.

Raghavendra, A.S., Gonugunta, V.K., Christmann, A., and Grill, E. (2010). ABA perception and signalling. *Trends in Plant Science* 15: 395–401.

Rai, K.K., Rai, N., and Rai, S.P. (2018). Salicylic acid and nitric oxide alleviate high temperature induced oxidative damage in *Lablab purpureus* L. plants by regulating biophysical processes and DNA methylation. *Plant Physiology and Biochemistry* https://doi.org/10.1016/j.plaphy.2018.04.023.

Rajashekar, C.B. (2000). Cold response and freezing tolerance in plants. In: *Plant–environment interactions*, 2nd edition (ed. R.E. Wilkinson), 321–341. New York, USA: Marcel Dekker, Inc.

Reddy, K.R., Davidonis, G.H., Johnson, A.S., and Vinyard, B.T. (1999). Temperature regime and carbon dioxide enrichment alter cotton boll development and fiber properties. *Agronomy Journal* 91: 851–858. https://doi.org/10.2134/agronj1999.915851x.

Rhodes, D. and Hanson, A.D. (1993). Quaternary ammonium and tertiary sulfonium compounds in higher plants. *Annual Review of Plant Physiology* 44: 357–384.

Rieu, I., Twell, D., and Firon, N. (2017). Pollen development at high temperature: from acclimation to collapse. *Plant Physiology* 173 (4): 1967–1976. https://doi.org/10.1104/pp.16.01644.

Rizhsky, L., Liang, H., Shuman, J. et al. (2004). When defense pathways collide. The response of Arabidopsis to a combination of drought and heat stress. *Plant Physiology* 134: 1683–1696.

Robertson, A.J., Reaney, M.J., Wilen, R.W. et al. (1994). Effects of abscisic acid metabolites and analogs on freezing tolerance and gene expression in bromegrass (*Bromus inermis* Leyss) cell cultures. *Plant Physiology* 105 (3): 823–830.

Rousch, J.M., Bingham, S.E., and Sommerfeld, M.R. (2004). Protein expression during heat stress in thermo-intolerant and thermo-tolerant diatoms. *Journal of Experimental Marine Biology and Ecology* 306: 231–243.

Routaboul, J.M., David, L.C., Dechorgnat, J. et al. (2014). Proanthocyanidin oxidation of *Arabidopsis* seeds is altered in mutant of the high-affinity nitrate transporter NRT2. 7. *Journal of Experimental Botany* 65 (3): 885–893.

Sakamoto, A. and Murata, N. (2002). The role of glycine betaine in the protection of plants from stress: clues from transgenic plants. *Plant, Cell & Environment* 25: 163–172.

Salvucci, M.E. and Crafts-Brandner, S.J. (2004). Inhibition of photosynthesis by heat stress: the activation state of Rubisco as a limiting factor in photosynthesis. *Physiologia Plantarum* 120: 179–186. https://doi.org/10.1111/ppl.2004.120.issue-2.

Savchenko, G., Klyuchareva, E., Abramchik, L., and Serdyuchenko, E. (2002). Effect of periodic heat shock on the inner membrane system of etioplasts. *Russian Journal of Plant Physiology* 49: 349–359. https://doi.org/10.1023/A:1015592902659.

Savicka M, & Škute N (2010). Effects of high temperature on malondialdehyde content, superoxide production and growth changes in wheat seedlings (*Triticum aestivum* L.).

Scarf, K.D., Berberich, T., Ebersberger, I., and Novel, L. (2012). The plant heat stress transcription factor (Hsf) family: structure, function and evolution. *Biochimica et Biophysica Acta* 1819: 104–119.

Sehgal, A., Sita, K., Bhandari, K. et al. (2018). Influence of drought and heat stress, applied independently or in combination during seed development, on qualitative and quantitative aspects of seeds of lentil (*Lens culinaris* Medikus) genotypes, differing in drought sensitivity. *Plant, Cell & Environment* https://doi.org/10.1111/pce.13328.

Shahbaz, M., Masood, Y., Perveen, S., and Ashraf, M. (2011). Is foliar-applied glycinebetaine effective in mitigating the adverse effects of drought stress on wheat (*Triticum aestivum* L.). *Journal of Applied Botany and Food Quality* 84: 192–199.

Sharma, P. and Dubey, R.S. (2005). Drought induces oxidative stress and enhances the activities of antioxidant enzymes in growing rice seedlings. *Plant Growth Regulation* 46 (3): 209–221.

Shi, J., Chen, Y., Xu, Y. et al. (2017). Differential proteomic analysis by iTRAQ reveals the mechanism of *Pyropia haitanensis* responding to high temperature stress. *Scientific Reports* 7 (1) https://doi.org/10.1038/srep44734.

Shi, Y., Zhang, J., Li, H. et al. (2018). Butanediol-enhanced heat tolerance in *Agrostis stolonifera* in association with alteration in stress-related gene expression and metabolic profiles. *Environmental and Experimental Botany* 153: 209–217.

Siddiqui, M.H., Al-Khaishany, M.Y., Al-Qutami, M.A. et al. (2015). Morphological and physiological characterization of different genotypes of faba bean under heat stress. *Saudi Journal of Biological Sciences* 22 (5): 656–663.

Silva, R.C.B., Araujo, M.N., Ornellas, F.L.S., and Dantas, B.F. (2018). Thermal stress and physiological changes in watermelon seeds. *Pesquisa Agropecuária Tropical* 48 (1): 66–74. https://doi.org/10.1590/1983-40632018v4850404.

Sinha, A.K., Hofmann, M.G., Römer, U. et al. (2002). Metabolizable and non-metabolizable sugars activate different signal transduction pathways in tomato. *Plant Physiology* 128 (4): 1480–1489.

Smirnoff, N. and Cumbes, Q.J. (1989). Hydroxyl radical scavenging activity of compatible solutes. *Phytochemistry* 28: 1057–1060.

Snider, J.L., Oosterhuis, D.M., Loka, D.A., and Kawakami, E.M. (2011). High temperature limits *in vivo* pollen tube growth rates by altering diurnal carbohydrate balance infield-grown *Gossypium hirsutum* pistils. *Journal of Plant Physiology* 168: 1168–1175. https://doi.org/10.1016/j.jplph.2010.12.011.

Song, X.S., Hu, W.H., Mao, W.H. et al. (2005). Response of ascorbate peroxidase isoenzymes and ascorbate regeneration system to abiotic stresses in *Cucumis sativus* L. *Plant Physiology and Biochemistry* 43: 1082–1088.

Song, L., Ding, W., Shen, J. et al. (2008, 2008). Nitric oxide mediates abscisic acid induced thermotolerance in the calluses from two ecotypes of reed under heat stress. *Plant Science* 175: 826–832. http://dx.doi.org/10.1016/j.plantsci.2008.08.005.

Song, G., Zhang, R., Zhang, S. et al. (2017). Response of microRNAs to cold treatment in the young spikes of common wheat. *BMC Genomics* 18: 1–15.

Spicher, L., Glauser, G., and Kessler, F. (2016). Lipid antioxidant and galactolipid remodeling under temperature stress in tomato plants. *Frontiers in Plant Science* 7: 167.

Spollen, W.G. and Nelson, C.J. (1994). Response of fructan to water deficit in growing leaves of tall fescue. *Plant Physiology* 106 (1): 329–336.

Stief, A., Simone, A., Hoffmann, K. et al. (2014). *Arabidopsis* miR156 regulates tolerance to recurring environmental stress through SPL transcription factors. *Plant Cell* 26: 1792–1807.

Stone, P. (2001). The effects of heat stress on cereal yield and quality. In: *Crop Responses and Adaptations to Temperature Stress* (ed. A.S. Basra), 243–291. New York: Food Products Press.

Sudhakar, C., Lakshmi, A., and Surabhi, G.K. (2001). Changes in the antioxidant enzyme efficacy in two high yielding genotypes of mulberry (*Morus alba* L.) under NaCl salinity. *Plant Science* 16: 613–619.

Surabhi, G.K., Madhusudhan, K.V., Sreenivasulu, N., and Sudhakar, C. (2000). Stress responses in two genotypes of mulberry (*Morus alba* L.) under NaCl salinity. *Indian Journal of Experimental Biology* 38: 192–195.

Surabhi, G.K., Reddy, A.M., and Sudhakar, C. (2003). NaCl effects on proline metabolism in two high yielding genotypes of mulberry (*Morus alba* L.) with contrasting salt tolerance. *Plant Science* 165: 1245–1251.

Suzuki, N. and Katano, K. (2018). Coordination between ROS regulatory systems and other pathways under heat stress and pathogen attack. *Frontiers in Plant Science* 9: 490. https://doi.org/10.3389/fpls.2018.00490.

Suzuki, N., Rivero, R.M., Shulaev, V. et al. (2014). Abiotic and biotic stress combinations. *The New Phytologist* 203: 32–43. https://doi.org/10.1111/nph.12797.

Tang, T., Liu, P., Zheng, G., and Li, W. (2016). Two phases of response to longterm moderate heat: variation in thermotolerance between Arabidopsis thaliana and its relative *Arabis paniculata*. *Phytochemistry* 122: 81–90. https://doi.org/10.1016/j.phytochem.2016.01.003.

Teplova, I.R., Farkhutdinov, R.G., Mitrichenko, A.N. et al. (2000). Response of tobacco plants transformed with the *ipt* gene to elevated temperature. *Russian Journal of Plant Physiology* 47 (3): 367–369.

Tong-Xiang, L., Zong-Shen, Z., Jian-Bo, W., and Rong-Qian, L. (2009). Changes in abscisic acid immunolocalization in heat-stressed pepper seedlings. *Pakistan Journal of Botany* 41: 1173–1178.

Tripp, J., Mishra, S.K., and Scharf, K.D. (2009). Functional dissection of the cytosolic chaperone network in tomato mesophyll protoplasts. *Plant, Cell & Environment* 32: 123–133.

Tubiello, F.N., Soussana, J., and Howden, S.M. (2007). Crop and pasture response to climate change. *Proceedings of the National Academy of Sciences of the United States of America* 104: 19686–19690.

Urano, K., Maruyama, K., Ogata, Y. et al. (2009). Characterization of the ABA-regulated global responses to dehydration in *Arabidopsis* by metabolomics. *The Plant Journal* 57: 1065–1078.

Vollenweider, P. and Günthardt-Goerg, M.S. (2005). Diagnosis of abiotic and biotic stress factors using the visible symptoms in foliage. *Environmental Pollution* 137: 455–465. https://doi.org/10.1016/j.envpol.2005.01.032.

Wahid, A. and Close, T.J. (2007). Expression of dehydrins under heat stress and their relationship with water relations of sugarcane leaves. *Biologia Plantarum* 51: 104–109. https://doi.org/10.1007/s10535-007-0021-0.

Wahid, A. and Ghazanfar, A. (2006). Possible involvement of some secondary metabolites in salt tolerance of sugarcane. *Journal of Plant Physiology* 163: 723–730.

Wallis, J.G. and Browse, J. (2002). Mutants of Arabidopsis reveal many roles for membrane lipids. *Progress in Lipid Research* 41: 254–278. https://doi.org/10.1016/S0163-7827(01)00027-3.

Wang, W., Vinocur, B., Shoseyov, O., and Altman, A. (2004). Role of plant heat shock proteins and molecular chaperones in the abiotic stress response. *Trends in Plant Science* 9: 244–252.

Wang, J., Gan, Y.T., Clarke, F., and McDonald, C.L. (2006). Response of chickpea yield to high temperature stress during reproductive development. *Crop Science* 46: 2171–2178. https://doi.org/10.2135/cropsci2006.02.0092.

Wang, Q.B., Xu, W., Xue, Q.Z., and Su, W. (2010). Transgenic *Brassica chinensis* plants expressing a bacterial *codA* gene exhibit enhanced tolerance to extreme temperature and high salinity. *Journal of Zhejiang University. Science* 11: 851–861.

Wang, G.F., Li, W.Q., Li, W.Y. et al. (2013). Characterization of rice NADPH oxidase genes and their expression under various environmental conditions. *International Journal of Molecular Sciences* 14: 9440–9458. https://doi.org/10.3390/ijms14059440.

Wang, X., Cai, J., Liu, F. et al. (2014). Multiple heat priming enhances thermo-tolerance to a later high temperature stress via improving subcellular antioxidant activities in wheat seedlings. *Plant Physiology and Biochemistry* 74: 185–192.

Wang, W., Teng, F., Lin, Y. et al. (2018). Transcriptomic study to understand thermal adaptation in a high temperature-tolerant strain of *Pyropia haitanensis*. *PLoS One* 13 (4): e0195842.

Wu, J., Wang, L., and Wang, S. (2017). microRNAs associated with drought response in the pulse crop common bean (*Phaseolus vulgaris* L.). *Gene* 10: 78–86.

Wu, Z., Liang, J., Zhang, S. et al. (2018). A canonical DREB2-type transcription factor in lily is post-translationally regulated and mediates heat stress response. *Frontiers in Plant Science* 9: 1–16.

Wyn Jones, R.W. and McDonnell, E. (1988). Glycinebetaine biosynthesis and accumulation in unstressed and salt-stressed wheat. *Journal of Experimental Botany* 39 (4): 421–430.

Xiong, L. and Zhu, J.K. (2001). Abiotic stress signal transduction in plants: molecular and genetic perspectives. *Physiologia Plantarum* 112 (2): 152–166.

Xu, X., Chen, C., Fan, B., and Chen, Z. (2006). Physical and functional interactions between pathogen-induced Arabidopsis WRKY18, WRKY40, and WRKY60 transcription factors. *The Plant Cell* 18: 1310–1326. https://doi.org/10.1105/tpc.105.037523.

Xu, Y., Wang, J., Bonos, S.A. et al. (2018). Candidate genes and molecular markers correlated to physiological traits for heat tolerance in fine fescue cultivars. *International Journal of Molecular Sciences* 19: 116. https://doi.org/10.3390/ijms19010116.

Yabuta, Y. (2016). Functions of heat shock transcription factors involved in response to photooxidative stresses in *Arabidopsis*. *Bioscience Biotechnology and Biochemistry* 80 (7): 1254–1263.

Yang, X., Liang, Z., and Lu, C. (2005). Genetic engineering of the biosynthesis of glycinebetaine enhances photosynthesis against high temperature stress in transgenic tobacco plants. *Plant Physiology* 138: 2299–2309.

Yu, H.D., Yang, X.F., Chen, S.T. et al. (2012). Down-regulation of chloroplast RPS1 negatively modulates nuclear heat responsive expression of HsfA2 and its target genes in Arabidopsis. *PLoS Genetics* 8: e1002669. https://doi.org/10.1371/journal.pgen.1002669.

Zandalinas, S.I., Rivero, R.M., Martínez, V. et al. (2016). Tolerance of citrus plants to the combination of high temperatures and drought is associated to the increase in transpiration modulated by a reduction in abscisic acid levels. *BMC Plant Biology* 16 (1): 105.

Zandalinas, S.I., Balfagón, D., Arbona, V., and Gómez-Cadenas, A. (2017). Regulation of citrus responses to the combined action of drought and high temperatures depends on the severity of water deprivation. *Physiologia Plantarum* 162 (4): 427–438. https://doi.org/10.1111/ppl.12643.

Zandalinas, S.I., Mittlerb, R., Balfagóna, D. et al. (2018). Plant adaptations to the combination of drought and high temperatures. *Physiologia Plantarum* 162: 2–12.

Zhang, Y.C., Yu, Y., Wang, C.Y. et al. (2013). Overexpression of microRNA OsmiR397 improves rice yield by increasing grain size and promoting panicle branching. *Nature Biotechnology* 31: 848–852. https://doi.org/10.1038/nbt.2646.

Zhao, Y., Aspinall, D., and Paleg, L.G. (1992). Protection of membrane integrity in *Medicago saliva* L. by glycinebetaine against the effects of freezing. *Journal of Plant Physiology* 140: 541–543.

Zhao, Y., Tian, X., Wang, F. et al. (2017). Characterization of wheat *MYB* genes responsive to high temperatures. *BMC Plant Biology* 17: 1–14.

Zinn, K.E., Tunc-Ozdemir, M., and Harper, J.F. (2010). Temperature stress and plant sexual reproduction: uncovering the weakest links. *Journal of Experimental Botany* 61: 1959–1968. https://doi.org/10.1093/jxb/erq053.

Index

Heat Stress Tolerance in Plants: Physiological, Molecular and Genetic Perspectives, First Edition.
Edited by Shabir Hussain Wani and Vinay Kumar.

D

E

F

G

Q

R

U

V

W

X

Y

www.ingramcontent.com/pod-product-compliance
Lightning Source LLC
LaVergne TN
LVHW011511200926
848484LV00001B/13